Springer Proceedings in Physics 9

Springer Proceedings in Physics

Managing Editor: H. K. V. Lotsch

Volume 1 *Fluctuations and Sensitivity in Nonequilibrium Systems*
Editors: W. Horsthemke and D. K. Kondepudi

Volume 2 *EXAFS and Near Edge Structure III*
Editors: K. O. Hodgson, B. Hedman, and J. E. Penner-Hahn

Volume 3 *Nonlinear Phenomena in Physics*
Editor: F. Claro

Volume 4 *Time-Resolved Vibrational Spectroscopy*
Editors: A. Laubereau and M. Stockburger

Volume 5 *Physics of Finely Divided Matter*
Editors: N. Boccara and M. Daoud

Volume 6 *Aerogels*
Editor: J. Fricke

Volume 7 *Nonlinear Optics: Materials and Devices*
Editors: C. Flytzanis and J. L. Oudar

Volume 8 *Optical Bistability III*
Editors: H. M. Gibbs, P. Mandel, N. Peyghambarian, and S. D. Smith

Volume 9 *Ion Formation from Organic Solids (IFOS III)*
Editor: A. Benninghoven

Volume 10 *Atomic Transport and Defects in Metals by Neutron Scattering*
Editors: C. Janot, W. Petry, D. Richter, and T. Springer

Springer Proceedings in Physics is a new series dedicated to the publication of conference proceedings. Each volume is produced on the basis of camera-ready manuscripts prepared by conference contributors. In this way, publication can be achieved very soon after the conference and costs are kept low; the quality of visual presentation is, nevertheless, very high. We believe that such a series is preferable to the method of publishing conference proceedings in journals, where the typesetting requires time and considerable expense, and results in a longer publication period. Springer Proceedings in Physics can be considered as a journal in every other way: it should be cited in publications of research papers as *Springer Proc. Phys.*, followed by the respective volume number, page number and year.

Ion Formation from Organic Solids (IFOS III)

Mass Spectrometry of Involatile Material

Proceedings of the Third International Conference
Münster, Fed. Rep. of Germany, September 16–18, 1985

Editor: A. Benninghoven

With 171 Figures

Springer-Verlag
Berlin Heidelberg New York Tokyo

Professor Dr. Alfred Benninghoven
Physikalisches Institut der Universität Münster, Domagkstraße 75
D-4400 Münster, Fed. Rep. of Germany

ISBN 3-540-16258-5 Springer-Verlag Berlin Heidelberg New York Tokyo
ISBN 0-387-16258-5 Springer-Verlag New York Heidelberg Berlin Tokyo

This work is subject to copyright. All rights are reserved, whether the whole or part of the material is concerned, specifically those of translation, reprinting, reuse of illustrations, broadcasting, reproduction by photocopying machine or similar means, and storage in data banks. Under § 54 of the German Copyright Law where copies are made for other than private use, a fee is payable to "Verwertungsgesellschaft Wort", Munich.

© Springer-Verlag Berlin Heidelberg 1986
Printed in Germany

The use of registered names, trademarks, etc. in this publication does not imply, even in the absence of a specific statement, that such names are exempt from the relevant protective laws and regulations and therefore free for general use.

Offset printing: Weihert-Druck GmbH, D-6100 Darmstadt
Bookbinding: J. Schäffer OHG, D-6718 Grünstadt
2153/3150-543210

Preface

The 3rd International Conference on Ion Formation from Organic Solids (IFOS III) was held at the University of Münster, September 16–18, 1985. The conference was attended by 60 invited scientists from all over the world. Of the 43 papers which were presented, 40 are included in these proceedings.

The aim of IFOS III was to promote the exchange of results and new ideas between scientists actively working in the field of mass spectrometry of involatile materials. Various aspects of the ion formation process – realization and optimization, theoretical understanding and analytical application – were treated, as well as instrumental developments. Some emphasis was placed on recent developments in time-of-flight and Fourier transform ion cyclotron resonance mass spectrometry, and its impact on the mass spectrometry of involatile materials.

The most important goal of the conference was to combine facets of the understanding of the most complex ion formation processes with the many different aspects of its analytical application. The participants came from a wide variety of different fields, including pure and applied physics and chemistry, medicine, pharmacy, and space research.

Finally, on behalf of all the conference participants, I would like to thank Dr. W. Sichtermann and Miss I. Bekemeier for the perfect preparation and technical organization of the conference.

The next conference in this series, IFOS IV, is planned for the autumn of 1987, in Münster.

Münster, January 1986 *A. Benninghoven*

Contents

Part I	^{252}Cf-Plasma Desorption

Use of Polymer Surfaces for Molecular Ion Adsorption and Desorption. By R.D. Macfarlane, C.J. McNeal, and R.G. Phelps 2

Electronic Sputtering of Biomolecules
By B. Sundqvist, A. Hedin, P. Håkansson, G. Jonsson, M. Salehpour, G. Säve, S. Widdiyasekera, and P. Roepstorff (With 4 Figures) 6

On the Charge-State Dependence of Secondary Ion Emission from Phenylalanine. By O. Becker, W. Guthier, K. Wien, S. Della-Negra, Y. le Beyec, and J.R. Cotter (With 5 Figures) 11

Particle Desorption from Non-Metallic Surfaces by High Energy Heavy Ions. By W. Guthier (With 6 Figures) 17

^{252}Cf-PDMS: Multiplicity of Desorbed Ions and Correlation Effects
By L. Schmidt and H. Jungclas (With 6 Figures) 22

Part II	Secondary Ion Mass Spectrometry (SIMS)

Surface Organic Reactions Induced by Ion Bombardment
By R.G. Cooks, B.-H. Hsu, W.B. Emary, and W.K. Fife
(With 1 Figure) 28

Ion Bombardment MS: A Sensitive Probe of Chemical Reactions Occurring at the Surface of Organic Solids
By B.T. Chait (With 1 Figure) 34

Ion-Neutral Correlations Following Metastable Decay
By K.G. Standing, W. Ens, R. Beavis, G. Bolbach, D. Main, B. Schueler, and J.B. Westmore (With 5 Figures) 37

Metastable Ion Studies with a ^{252}Cf Time-of-Flight Mass Spectrometer. By S. Della-Negra and Y. le Beyec (With 3 Figures) .. 42

Increasing Secondary Ion Yields: Derivatization/SIMS
By D.A. Kidwell, M.M. Ross, and R.J. Colton (With 6 Figures) 46

Aspects and Applications of Derivatization/SIMS
By M.M. Ross, J.E. Campana, R.J. Colton, and D.A. Kidwell
(With 7 Figures) 51

Influence of the Target Preparation on the SI-Emission of Organic
Molecules. By A. Eicke and A. Benninghoven (With 6 Figures) 56

Secondary Ion Formation Processes in Amino Acid–Metal
Adsorption Systems. By D. Holtkamp, M. Kempken, P. Klüsener,
and A. Benninghoven (With 5 Figures) 62

Analytical Applications of High-Performance TOF-SIMS
By W. Lange, D. Greifendorf, D. van Leyen, E. Niehuis,
and A. Benninghoven (With 4 Figures) 67

TOF-SIMS of Polymers in the High Mass Range
By I.V. Bletsos, D.M. Hercules, D. van Leyen, E. Niehuis,
and A. Benninghoven (With 4 Figures) 74

The Application of Time-of-Flight Secondary Ion Mass Spectrometry
in the Characterization of Apolipoprotein Mutants
By H.-U. Jabs, M. Walter, G. Assmann, and A. Benninghoven
(With 4 Figures) 79

Part III   Liquid SIMS Including FAB

Sputtering Yields from Liquid Organic Matrices
By D.F. Barofsky and E. Barofsky (With 4 Figures) 86

Sputtering from Liquid and Solid Organic Matrices
By S.S. Wong, K.P. Wirth, and F.W. Röllgen (With 3 Figures) 91

Secondary Ion Emission from Glycerol and Silver Supported Organic
Molecules. By M. Junack, W. Sichtermann, and A. Benninghoven
(With 7 Figures) 96

Temperature Effects in Particle Bombardment Mass Spectrometry of
Methanol. By R.N. Katz, T. Chaudhary, and F.H. Field 102

Internal Energy Distribution of Ions Emitted in Secondary Ion Mass
Spectrometry. By E. de Pauw, G. Pelzer, J. Marien, and P. Natalis
(With 3 Figures) 103

Fast Atom Bombardment of Peptides Above 5000 Daltons
By C. Fenselau and K. Hyver (With 4 Figures) 109

Amino Acid Sequencing of Peptide Mixture: Structural Analysis of
Human Hemoglobin Variants (Digit Printing Method)
By T. Matsuo, T. Sakurai, I. Katakuse, H. Matsuda, Y. Wada,
and A. Hayashi (With 4 Figures) 113

Oligonucleotide Sputtering from Liquid Matrices
By L. Grotjahn (With 4 Figures) 118

Some Experiments on the Production of Ions in Soft Ionisation Mass
Spectrometry. By D. Renner and G. Spiteller (With 9 Figures) 126

Decompositions Occurring Remote from the Charge Site: A New
Class of Fragmentation of FAB-Desorbed Ions
By K.B. Tomer and M.L. Gross (With 5 Figures) 134

Part IV Laser-Induced Ion Formation

Laser and Plasma Desorption: Matrices and Metastables in Time-of-Flight Mass Spectrometry. By R.J. Cotter, J. Honovich, J. Olthoff,
P. Demirev, and M. Alaim (With 7 Figures) 142

Evidence for Simultaneous Generation of Ion Pairs in Laser Mass
Spectrometry. By D.M. Hercules (With 1 Figure) 147

The Influence of the Substrate on Ultraviolet Laser Desorption Mass
Spectrometry of Biomolecules. By F. Hillenkamp, D. Holtkamp,
M. Karas, and P. Klüsener (With 5 Figures) 153

On Different Desorption Modes in LDMS
By B. Lindner and U. Seydel (With 6 Figures) 158

Part V Other Ion Formation Processes

"Spontaneous" Desorption of Negative Ions from Organic Solids and
Films of Ice at Low Temperature. By S. Della-Negra, C. Deprun,
Y. le Beyec, J. Benit, J.P. Bibring, and F. Rocard (With 6 Figures) .. 164

Electric Pulse-Induced Desorption Compared to Other Techniques -
Mechanism, Mass Spectra, and Applications
By F.J. Mayer, F.R. Krueger, and J. Kissel (With 3 Figures) 169

Part VI Instrumentation

A New Dual-MS Technique Combining Negative Ion Formation
by Plasma Desorption with EI-like Positive Ion Formation by
In-Beam Desorption. By H. Brandenberger and F.B.Ch. West
(With 5 Figures) 174

The Chemical Ionization/Particle-Induced Ion Source
By R.B. Freas and J.E. Campana (With 3 Figures) 179

Design of Modern Time-of-Flight Mass Spectrometers
By H. Wollnik (With 4 Figures) 184

Design of an Organic SIMS Instrument with Separate Triple Stage
Quadrupole (TSQ) and Time-of-Flight (TOF) Spectrometers
By B.L. Bentz and R.E. Honig (With 3 Figures) 192

High-Resolution TOF Secondary Ion Mass Spectrometer
By E. Niehuis, T. Heller, H. Feld, and A. Benninghoven
(With 5 Figures) 198

Part VII Fourier Transform Ion Cyclotron Resonance

Laser Desorption Fourier Transform Mass Spectrometry: Mechanisms
of Desorption and Analytical Applications
By M.P. Chiarelli, D.A. McCrery, and M.L. Gross (With 2 Figures) .. 204

Desorption Ionization and Fourier Transform Mass Spectrometry for
the Analysis of Large Biomolecules
By D.H. Russell and M.E. Castro (With 3 Figures) 209

Application of Secondary Ion Mass Spectrometry Combined
with Fourier Transform Ion Cyclotron Resonance. By S. Plesko,
P. Grossmann, M. Allemann, and H.P. Kellerhals (With 8 Figures) ... 213

Index of Contributors 219

Part I

^{252}Cf-Plasma Desorption

Use of Polymer Surfaces for Molecular Ion Adsorption and Desorption

R.D. Macfarlane, C.J. McNeal, and R.G. Phelps

Texas A & M University, Department of Chemistry, College Station, TX 77843, USA

1. Introduction The desorption of molecules and molecular ions from surfaces as a consequence of excitation by an incident fast or slow particle is a classic example of a good news-bad news scenario. The good news is that organic mass spectrometry has experienced a revolution in development and application. The bad news is that the understanding of the processes involved is still in a state of chaos and there is no evidence that the atmosphere has improved to a level where an orderly evolution of concepts will emerge. Part of the problem is emotional, linked to the personalities and ambitions of those, like ourselves, who are actively promoting a particular variant and model, but another part is related to the complexity of the problem. Unlike studies of atomic and molecular ion emission using highly disciplined surfaces where theory and experiment are sharpened to elegant simplicity, the fragile and reactive nature of the organic molecule makes the study of the emission of their molecular ions a dirty business. One aspect of the high vacuum technology that has made possible a study of adsorbate-substrate interaction is the use of Langmuir adsorption at a solid-gas interface. By careful control of pressure and exposure time, it is possible to lay down adsorbate layers with varying degrees of surface coverage and to study the influence of surface concentration on the desorption-ionization process [1]. It is also possible to perform Langmuir adsorption experiments at a solid-liquid interface. In these studies, solution concentration plays the role of partial pressure in controlling the surface concentration of adsorbate. Rapid removal of the liquid phase after equilibrium has been reached leaves a surface containing adsorbate in a particular state dictated by the conditions of the thermodynamics of adsorption for that particular system. The adsorbate can then be analyzed by a surface sensitive mass spectrometric method. While the adsorbate-substrate system is not as well defined as in gas-solid adsorption, it is a considerable improvement in terms of control over most of the methods used to prepare organic deposits such as electrospraying or solution evaporation. The variables available include solution concentration and temperature, which can yield thermodynamic state function data for adsorption, different surfaces and surface modification to achieve specificity for adsorption, and the influence of the composition of the solution (pH, ionic strength, counter ion) in preparing the adsorbate in different forms (protonated, ion pair, molecular aggregates).

Our first studies in this approach utilized a cation exchange polymer, Nafion, as the substrate, and we were able to demonstrate the feasibility of the method using inorganic and organic cationic species [2]. In the work reported here, we have expanded this approach to include other polymer surfaces and a different modification that transforms the surface to an anion exchange medium [3]. The 252-Cf plasma desorption method was used for mass analysis of the adsorbed layers [4].

2. Adsorption on Polypropylene and Poly(ethylene terephthalate) The standard backing we have used for most of our studies in the past decade is aluminized poly(ethylene terephtalate) (Mylar). Samples of organic molecules are deposited on the metal side of the film and the conducting layer serves as a high voltage equipotential surface for accelerating ions from the surface during the mass analysis. The Mylar film is extremely thin (50 nm) which means that 252-Cf

fission fragments can pass through the film from the back side with minimal attenuation of the energy deposition characteristics of the fast (100 MeV) heavy ions. After our experience with adsorption on Nafion, we became interested in the general application of polymer surfaces as a matrix for molecular ion adsorption, and as a first step we reversed the aluminized Mylar film to determine whether we could obtain a meaningful mass spectrum from the polymer side that would not be distorted by surface charging effects or attenuation of the electric field by the dielectric medium between the Al and the desorption surface. We found that it was possible to obtain mass spectra with a quality very close to that observed using a metallized surface. For Mylar, a fragmentation pattern was observed that could be correlated with the structure of the polymer. Similar measurements using Al-polypropylene films showed that this film could also be used, but no mass spectrum was observed that could be attributed to polypropylene.

The original plan was to use Mylar as a base polymer and to coat the surface with other polymeric materials containing different functional groups. In the first measurement, we carried out adsorption studies with untreated Mylar and polypropylene to establish a base for comparison when the surfaces were coated with other polymers. To our surprise, we found that both Mylar and polypropylene exhibited significant cation exchange behavior in the untreated form. Consequently, our intial study involved the study of the adsorption-desorption character of these polymers. For these studies, Al-polypropylene was used because of the possibility that for Mylar, the cation adsorption sites might be carboxyl groups formed by the cleavage of ester linkages in the polymer. But in the final analysis, there was no discernible difference in the performance of the two polymers in terms of specificity and capacity, and our conclusion was that the cation exchange behavior of these materials was due to impurities chemisorbed on the surface and not related to details of the polymer structure [5].

3. Adsorption of CsI The first solute used in the adsorption studies was CsI in aqueous and ethanol solution. Mass spectra of the adsorbed layer from this solution showed only the presence of Cs^+ ions. There was no evidence for I^- ion adsorption. This meant that the surface of the polymer was negatively charged. Subsequent studies on the origin of the negative charge pointed to the presence of strongly adsorbed Cl^- and OH^- ions on the surface presumably coming from impurities in the polymer and solution and from adsorbed H_2O. A study was made on the influence of solution concentration on the surface concentration of adsorbed species covering a concentration range from 10^{-5} M to 10^{-2} M. Plotting Cs^+ ion intensity (determined by 252-Cf-PDMS) vs. solution concentration, a Langmuir-type adsorption curve was obtained with saturation intensity achieved at 5×10^{-3} M for an ethanol solution. Because of the higher solubility in aqueous solutions, the Cs^+ ion adsorption equilibrium constant was not as large as for ethanol solutions. The 252-Cf-PD spectrum of adsorbed Cs^+ was quite different from an electrosprayed sample of CsI. There were no cluster ions at low concentration, and at high concentration the major cluster ion was Cs_2Cl^+ and not Cs_2I^+ which we have interpreted as indicative of the nature of the adsorption site.

4. Adsorption of Rhodamine 6-G Hydrochloride This species was selected as an example of an organic cation for the adsorption studies. The same type measurements were carried out for this species using aqueous solutions and results were obtained similar to what was observed for CsI adsorption: adsorption of only the cation, a Langmuir-type adsorption with saturation molecular intensity (determined by 252-Cf-PDMS) occurring at 10^{-4} M. Fluorescence and Beer-Lambert absorption at saturation gave evidence for full monolayer coverage at this concentration, which was corroborated by a diminution in the intensity of the peaks due to Mylar in the mass spectrum when this polymer was used. The molecular ion intensity at saturation was essentially the same as for a multilayer electrosprayed sample, indicating that for this species, molecular ions are emitted only from the surface layer. At higher surface coverage, dimer ions of Rhodamine 6-G containing the Cl^- counter ion were observed. These probably are formed in solution at the higher concentration and are adsorbed on the surface in this form.

In order to learn more about the nature of the adsorption interaction, competitive adsorption studies were carried out involving Rhodamine 6-G and CsI. Since both of these species are strongly adsorbed as separate solutes, the adsorption from a binary solution gave information on the competition of these two species for the negatively charged sites. If the site is a small compact region of negative charge, Cs^+ ion adsorption would be preferred because of the dominance of the electrostatic interaction. However, if the negative site has a diffuse charge density, then the combination of electrostatic and polarization forces would favor Rhodamine 6-G adsorption because of the polarizability of the aromatic rings in its structure. The results showed that Rhodamine adsorption dominated even when the molar ratio of Cs:Rhodamine was as high as 4:1. This means that the adsorption site resembles a large polarizable organic anion favoring the adsorption of organic cations even in the presence of high concentrations of alkali metal ions. In later studies, this property was utilized to selectively adsorb organic species from solution in the presence of high salt concentrations.

5. Adsorption of Amino Acids and Leu-enkephalin.

After the studies discussed above determined the nature of the adsorption behavior of Mylar and polypropylene, this method of sample preparation was applied to the study of important organic species that were not so strongly basic as Rhodamine as to exist predominantly as a cation in aqueous solution. The simplest of these is the ammonium ion (NH_4^+) which is strongly adsorbed on Nafion and which can be desorbed intact using the fission fragments of 252-Cf-PDMS. However, attempts to adsorb-desorb NH_4^+ ions on Mylar failed. This we attribute to the adsorption site behaving as a strong base capable of deprotonating the NH_4^+ ion. The sulfonate group on Nafion is, by contrast, a weak base (conjugate of a strong acid) so that its interaction with the NH_4^+ ion is essentially electrostatic. The same problem was encountered with simple amino acids, even those that are strongly basic (histidine, tryptophan) which in an acidic aqueous medium are positively charged. With the knowledge from the literature [6] that proteins readily adsorb on polymer surfaces, it became a matter of determining how large a peptide or protein had to be before adsorption took place. We next investigated the adsorption of leu-enkephalin from aqueous solution covering the concentration range previously used in the Rhodamine-Cs studies. From 10^{-5} M to 10^{-3} M there was no evidence for peptide adsorption. But above that concentration, adsorption-desorption of molecular ions was observed. At 10^{-3} M, the spectrum of molecular ions was more intense than that of a multilayer electrosprayed sample. There were also other differences. The dominant molecular ions were M±H ions of nearly equal intensity, and a very low level Na^+ intensity, orders of magnitude lower than the electrosprayed sample of the same material and, overall, a much cleaner spectrum free from peaks due to impurities. In the electrosprayed sample, molecular ions containing single and multiple Na^+ ions were observed. At the onset of the appearance of the molecular ions of this peptide, the dimer ion was also observed. This, coupled with the relatively high solution concentration required to effect adsorption, implies that it is the dimer that is being adsorbed and that the monomer ions are a consequence of the dissociation of a dimer ion during the fast ion-induced desorption process [7].

6. Conversion of Mylar to an Anion Exchange Surface [3]

We have now demonstrated that cations can be adsorbed and desorbed from Nafion and Mylar surfaces as a consequence of their negative surface charge. To expand the versatility of the method it would also be desirable to have available a positively charged surface for anion adsorption. We wished also to continue to use Mylar as the base polymer film because of its desirable mechanical properties. We have been able to reverse the polarity of the surface by a new kind of surface modification that involves "dissolving" a hydrophobic molecule into the surface that also contains a positively charged functional group. The molecule chosen for this modification was tridodecylmethyl ammonium chloride (TDMAC). The three dodecyl moieties are long hydrocarbon chains that embed into the Mylar surface when it is solvated with a non-polar solvent system such as toluene-petroleum ether. The molecule attaches strongly and is immobilized when in contact with an aqueous solution. The quaternary ammonium function extends into the polar phase and functions as an anion exchange site.

7. Adsorption of CsI on TDMAC-Modified Mylar   Aqueous solutions of CsI were exposed to a TDMAC-modified Mylar surface. The 252-Cf PD mass spectrum of the adsorbed layer showed a strong I^- ion in the negative ion spectrum plus cluster ions of I containing the TDMA moiety. The positive ion spectrum contained no Cs^+ ions but did show cluster ions comprised of I^- and the TDMA cation. Thus, by modification of the Mylar surface by the adsorption of a hydrophobic quaternary ammonium ion, the polarity of the surface was reversed and the surface was found to have a high capacity for anion adsorption.

8. Concluding Remarks   The use of polymer and surface modified polymer surfaces show great promise for selective adsorption and as a means of forming adsorbate layers with different surface concentrations, providing a means of controlling adsorbate states in a manner equivalent to what is used in gas-surface studies. The mass spectrum is a powerful and sensitive analysis for determining not only the kind of species adsorbed but also the surface concentration. In addition, the attenuation of the mass spectrum of the substrate is a confirmational signature for the degree of surface coverage. The versatility of the solid-liquid adsorption equilibrium should provide ample opportunity to develop better sample preparation methodologies.

Acknowledgements   The financial support of this research by the U.S. National Science Foundation (CHE-82-06030), the National Institutes of Health (GM-26096) and the Welch Foundation (Grant A-258) is gratefully acknowledged.

References

[1] M.L. Yu and N.D. Lang: Phys. Rev. Lett. 50, 127, (1983)
[2] E.A. Jordan, R.D. Macfarlane, C.R. Martin and C.J. McNeal: Int. J. Mass Spectrom. Ion Phys. 53, 345, (1983)
[3] "The Use of a Stationary Cationic Surfactant as a Selective Matrix in 252-Cf-Plasma Desorption Mass Spectrometry", C.J. McNeal and R.D. Macfarlane: (submitted to J. Am. Chem. Soc.)
[4] R.D. Macfarlane: Anal. Chem. 55, 1247A, (1983)
[5] "Mass Spectrometric Study of the Ion-Exchange Behavior of Polypropylene and Poly(ethylene terephthalate) Films", R.D. Macfarlane, C.J. McNeal and C.R. Martin: (submitted to Anal. Chem.)
[6] A.L. Iordanski, A. Ja Polischuk, and G.E. Zaikov: JMS-Rev. Macromol. Chem. Phys. C23, 33, (1983)
[7] "Mass Spectrometric Study of the Adsorption of Leu-Enkephalin on a Mylar Surface" R.D. Macfarlane and R.G. Phelps: (submitted to Langmuir)

Electronic Sputtering of Biomolecules

B. Sundqvist, A. Hedin, P. Håkansson, G. Jonsson, M. Salehpour, G. Säve, and S. Widdiyasekera

Tandem Accelerator Laboratory, Uppsala University, Box 533, S-751 21 Uppsala, Sweden

P. Roepstorff

Department of Molecular Biology, Odense University, DK-5230 Odense M, Denmark

1. Introduction

The erosion of surfaces under particle bombardment is called sputtering. For metallic targets, sputtering by elastic collisions between screened nuclei (nuclear sputtering) is the dominating mechanism. This effect is most important at low velocities, i.e. below the Bohr velocity, where nuclear stopping is the main energy loss mode. In insulators, where lifetimes of excited electronic states may be long enough to allow excitation energy to be transferred to atomic motion, sputtering due to electronic processes can occur as well. Sputtering of biomolecules by fast ions i.e. ions with a velocity larger than the Bohr velocity, has been shown to be related to electronic stopping [1,2] and may therefore be called electronic sputtering. Most of the ejected particles in sputtering processes are neutral. As recently shown at Uppsala by Salehpour et.al. [3] that is also the case for electronic sputtering of biomolecules,as will be discussed in some detail below. The analytical application of electronic sputtering of biomolecules is the method invented by Macfarlane and coworkers [4] called ^{252}Cf-Plasma Desorption Mass Spectrometry (PDMS). As fission fragments are fast ions, this method may be described as fast ion-solid sample - SIMS with the time-of-flight technique. One of the most promising developments in the field of PDMS involves the use of samples consisting of adsorbed monolayers or submonolayers of biomolecules on polymer backings [5,6]. In this report, sample application by adsorbing proteins on nitrocellulose is discussed.

2. Yields of intact neutral molecules in electronic sputtering of biomolecules

Almost all studies in the field of electronic sputtering of biomolecules have been concerned with the secondary ions ejected. In the theoretical description of the process the ionization step is a troublesome complication, and therefore experimental data on neutrals ejected are of great interest. Salehpour et.al. [3] have recently developed a method to measure the yield of intact neutral molecules ejected in sputtering of biomolecules. The sputtered particles are collected on a single crystal Si-wafer (~ 1 cm^2), positioned close to the production target (see fig. 1). A film of biomolecules prepared with the electrospray technique was bombarded with fast ions (90 MeV ^{127}I^{14+}) from the Uppsala EN-tandem accelerator. The collector was then analyzed in two ways. In the first method the collector was directly transferred in vacuum to a PDMS setup and a low intensity beam (1000 particles/sec) of 90 MeV ^{127}I^{16+} was used to make a PDMS-analysis of the collector surface. With the PDMS method relative yield measurements can be performed. The amount of molecules collected was always less than one monolayer and the sticking coefficient was assumed to be 1.

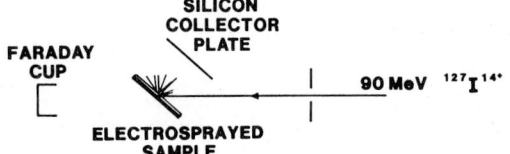

Fig. 1 The principle of the experimental setup used to measure sputtering-yields of intact neutral molecules.

The absolute yields were determined by performing amino acid analysis [7] of the collector. As one monolayer of an amino acid of 1 cm^2 corresponds to about one nanomole, the analysis is performed close to the detection limit of state of the art amino acid analysis. Using the methods described the amino acid leucine (MW 131) was studied with a 100 pA beam of 90 MeV ^{127}I^{14+}. If one assumes cosθ angular distribution for the ejected molecules, the intact neutral molecule yield was measured to be 1180±280 per incident ion [3]. As the yield of positive (protonated) molecular ions is about 0.1 the neutral to charge ratio for intact molecules is therefore about 10^4. Recently the first method was applied to the peptide LHRH (luteinizing hormone-releasing hormone, MW 1182). In fig 2a the PD-spectrum of positive ions from a Si-collector in that experiment is shown. The protonated molecule can be identified in the spectrum. Also shown is the long flight-time part of the spectrum from a clean collector surface. In fig 2b the PD spectrum of a spincasted LHRH [8] sample is shown. The spincasted film has an estimated thickness of 10Å, i.e. of the order of one monolayer as measured by ellipsometry. The same number of 90 MeV ^{127}I^{6+} was used to collect the three spectra. If one assumes that one monolayer is collected in the experiment and the angular distribution is cosθ the "sputter" yield of intact LHRH is about 500.

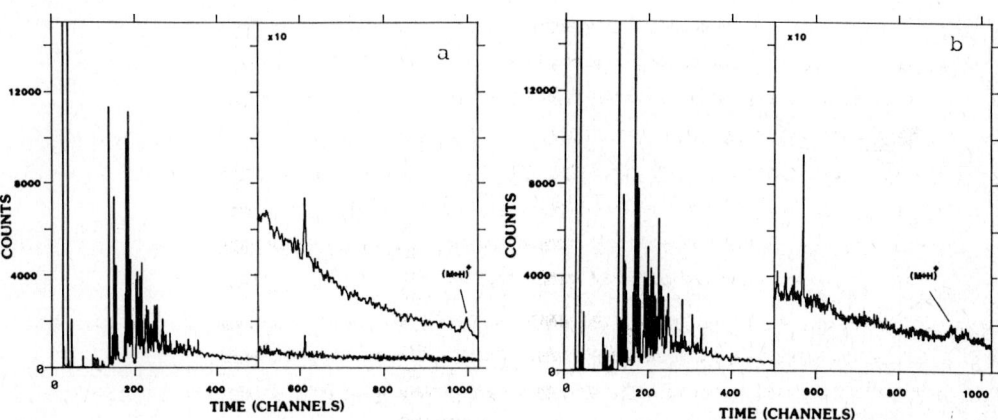

Fig. 2 ^{127}I-PDMS spectrum of positive ions from a Si-collector on which LHRH has been sputter-deposited. Dose: 5·10^{11} ^{127}I (a). The lower spectrum in fig 2a is the spectrum for a pure Si-backing. The corresponding spectrum for an LHRH-spin-casted sample (b).

3. Desorption and ionization of proteins and peptides adsorbed on nitrocellulose films

New methods of sample preparation for SIMS-analysis of biomolecules are needed. From a fundamental point of view the molecules should if possible be adsorbed on a well characterized backing in a well-defined surface state. This is far from the case in presently used methods, at least in PDMS-studies i.e. when the electrospray method is used. For molecular specific monolayer and submonolayer analysis purposes, the ultimate sensitivity can probably only be reached with such surface adsorption preparation procedures. Jordan et.al. [5,6] have used an ion conducting polymer, Nafion, as backing in PDMS-studies. In attempts to combine gel-electrophoresis and PDMS-analysis we have found that nitrocellulose is an attractive backing for adsorption-desorption-ionization of peptides and proteins [9]. A droplet of a molecule solution is applied to a nitrocellulose film on a metal backing and then the film is rinsed several times with a solution of 10 % acetic acid, 10 % methanol and 80 % water. The rinsing procedure was found to be very important in order to remove salt contaminants. It also seems that this removal of salts is associated with the production of molecular ions with less internal energy. However, the most striking effect in using nitrocellulose as backing is the enhanced multiply charged ion-production. In fig 3 this latter effect is illustrated by a comparison of ^{127}I-PDMS-spectra of positive ions of porcine trypsin from an electrosprayed sample and a sample prepared with the new method. As 20 kV acceleration voltage was used in the time-of-flight spectrometer, the 6^+ trypsin ions have an energy of 120 keV when they hit the stop detector. Using nitrocellulose as backing in PDMS-studies we have been able to get useful mass spectra of positive ions of a number of peptides and proteins which have failed to give mass spectra with electrosprayed samples. One

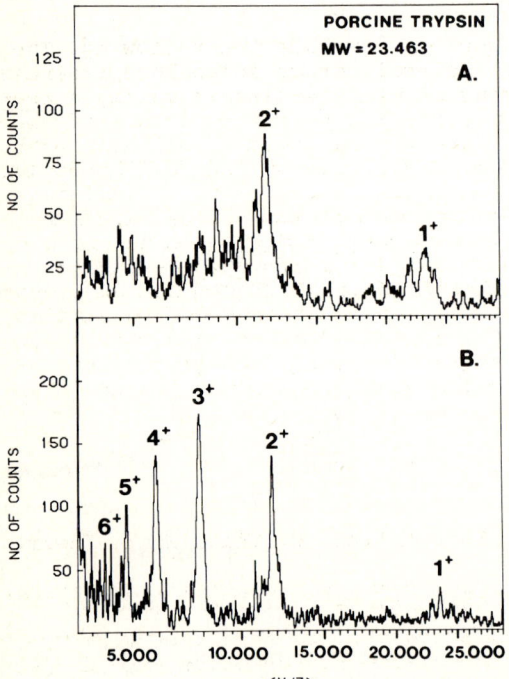

Fig. 3 ^{127}I-PDMS spectra of positive ions from porcine trypsin. Electrosprayed sample (upper spectrum) and sample prepared by adsorbing trypsin from a solution onto a nitrocellulose film (lower spectrum).

such example is an insect cuticle protein, the structure of which has been studied at Odense for the last few years [10]. However, sequence data and SDS-gel-electrophoresis have given quite different molecular weights, namely 15323 and 21600 respectively. With the spectrum in fig 4 it was possible to determine the molecular weight to 15320±100 and therefore solve the problem [11].

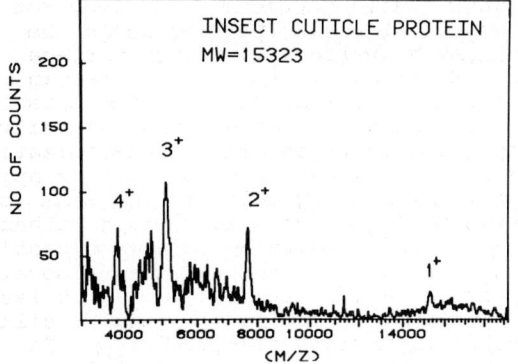

Fig. 4 ^{127}I-PDMS spectrum of positive ions from an insect cuticle protein [10].

4. Discussion and conclusions

Electronic sputtering of biomolecules is still a virgin field of research. The two projects discussed here i.e. the study of neutrals ejected and sputtering of biomolecular ions from layers of molecules adsorbed on polymer backings, have only been started recently at Uppsala and new experiments are under way.

However, already at this stage one can conclude the following. The experimental data on sputtering of neutral intact molecules presented here show that electronic sputtering of biomolecules is mainly a <u>neutral</u> ejection process. In the future the analytical potential of electronic sputtering may therefore be greatly improved if suitable postionization techniques are developed. The volume of "intact" molecules sputtered per incident ion indicates that a molecule in the molecular weight region 100000 may be possible to desorb intact. We have also demonstrated for the first time that large labile biomolecules can be "sputter"-deposited on a backing. The sample preparation technique discussed, i.e. the adsorption of proteins on nitrocellulose films, has important advantages over the main technique used at present in PDMS-studies i.e. the electrospray technique. Proteins can be adsorbed from water solutions i.e. in their native or denaturated state. Furthermore, the enhanced multiply charged ion production simplifies detection and peak identification of large molecular ions.

References

1. P. Dück, W. Treu, W. Galster, H. Fröhlich, H. Voit: Nucl. Instr. Meth. 168 (1980) 601
2. P. Håkansson, A. Johansson, I. Kamensky, B. Sundqvist, J. Fohlman, P. Peterson: IEEE Trans. Nucl. Sci. NS-28-2 (1981) 1776
3. M. Salehpour, P. Håkansson, B. Sundqvist. S. Widdiyasekera: Proc. from the 11th Int. Conf. Atomic Collisions in Solids, Washington D.C. 1985 (to appear in Nucl. Instr. Meth.)

4. D.F. Torgerson, R.P. Skowronski, R.D. Macfarlane: Biochem. Biophys. Res. Commun. 60 (1974) 616
5. E.A. Jordan, R.D. Macfarlane, C.R. Martin, C.J. McNeal: Int. J. Mass. Spectrom. Ion Phys. 53 (1983) 345
6. E.A. Jordan, C.R. Martin, C.J. McNeal, R.D. Macfarlane: Paper presented at the 33rd Annual Conf. on Mass. Spec. and Applied Topics, San Antonio, 1984
7. D.H. Spackman, W.H. Stein, S. Moore: Anal. Chem. 30 (1958) 1190
8. G. Säve, P. Håkansson, B. Sundqvist, R.E. Johnson, U. Jönsson: TLU 127/85, Tandem Laboratory Report, Uppsala, Sweden, 1985
9. G. Jonsson, A. Hedin, P. Håkansson, B. Sundqvist, G. Säve, P. Nielsen, P. Roepstorff, K.E. Johansson, I. Kamensky, M. Lindberg: TLU 125/85, Tandem Laboratory Report, Uppsala, Sweden, 1985 (submitted for publication in Anal. Chem.)
10. P. Højrup, P. Roepstorff, S.O. Andersen: (submitted to Biochem. J.)
11. P. Roepstorff, B. Sundqvist, P. Højrup, G. Jonsson, P. Håkansson, S.O. Andersen: (manuscript in preparation)

On the Charge-State Dependence of Secondary Ion Emission from Phenylalanine

O. Becker, W. Guthier, and K. Wien

Institut für Kernphysik, Technische Hochschule, D-6100 Darmstadt, F.R.G.

S. Della-Negra, Y. le Beyec, and J.R. Cotter

Institut de Physique Nucléaire, F-91406 Orsay, France

Thin films of phenylalanine evaporated onto Al foils were irradiated by MeV beams of N, Ne, S, Ar, Ca, Ni, Kr, Sn, W, Pb and U using the accelerator facilities of ALICE/Orsay and UNILAC/GSI-Darmstadt. By means of time-of-flight mass spectrometry the emission of secondary ions was investigated as function of the charge state q of the primary ion. Experiments show that the yields of ions correlated with molecular structure like $(M \pm 1)^{\pm}$ depend in addition strongly on the equilibrium charge inside the solid. These findings lead to conclusions about the time and the area of emission relative to the primary ions impact.

1 Introduction

When high energetic heavy ions penetrate a solid, their original charge state is rapidly altered by electron-loss or -capture processes in layers underneath the surface, until an equilibrium charge-state distribution is reached. Secondary ion emission induced by the heavy ion impact occurs from the surface, but recent experiments have shown [1,2,3,4] that the charge changing inside the solid has an influence on this surface process. In this work, we exhibit the role of the incident charge q as well as of the mean equilibrium charge q_{eq} for secondary ion emission. The intention is to find out if secondary ions can be used as spectators of the equilibrium charge state distribution inside a solid. This quantity is an important ingredient to theoretical and semi-empirical treatments of the electronic energy loss and related phenomena.

2 Experimental Methods

As in a recent investigation [4] the interest is again focussed on the aminoacid phenylalanine (=Phe). The compound was evaporated with a thickness of 1500 Å onto 2 μm-Al foils. The samples were irradiated in Orsay by Ne, Ar and Kr at a constant energy of 1.2 MeV/u and in Darmstadt by S, Ca, Ni, Kr, Sn, W, Pb and U at 1.4 MeV/u. The method of selecting definite charge states and the time-of-flight technique used to measure secondary ion yields have been described elsewhere [4].

The results presented in the following were obtained during several accelerator runs, where usually only two or three different beams were available. In Orsay the combination Ne-Ar-Kr could be used in one period, in Darmstadt N-U, S-Ca-Kr and W-Pb-U. During these periods always the same sample was irradiated with about 200 ions/s keeping the TOF instrumentation in the vacuum chamber permanently under low pressure (10^{-6} mbar). Even under these precautions slight changes of the samples due to sublimation (25 Å Phe per day) and to contamina-

---

This work is supported by the Deutsche Forschungsgemeinschaft

tion were observed. Yield ratios of positive and negative ions are not presented, because of changes in ion detection efficiency when the TOF system was switched from positive to negative ion acceptance.

3 Results and Discussion

As an example, yields of several secondary ions ejected by Ca are shown in Fig.1 as function of q. The yields are corrected for dead time losses [5] but not for detection efficiency. The curves demonstrate a behaviour typical for organic samples: The molecule specific ions, i.e. the parent ions $(M\pm 1)^{\pm}$, the cluster ions $(2M\pm 1)^{\pm}$ and fragment ions correlated to molecular structure, follow more or less bended curves (in log-log scale); their relative yields change little even if another kind of projectile is used (see also Fig.4). Small unspecific fragment ions like C^+ and C_2^-, however, have yields which decrease much steeper towards lower q values; the shape of their yield curves is describable by uniform exponents. Unspecific heavier fragment ions like the numerous hydrocarbon ions observed with U-beams have intermediate yield curves, with the tendency that the exponent decreases with increasing number of hydrogen atoms [4].

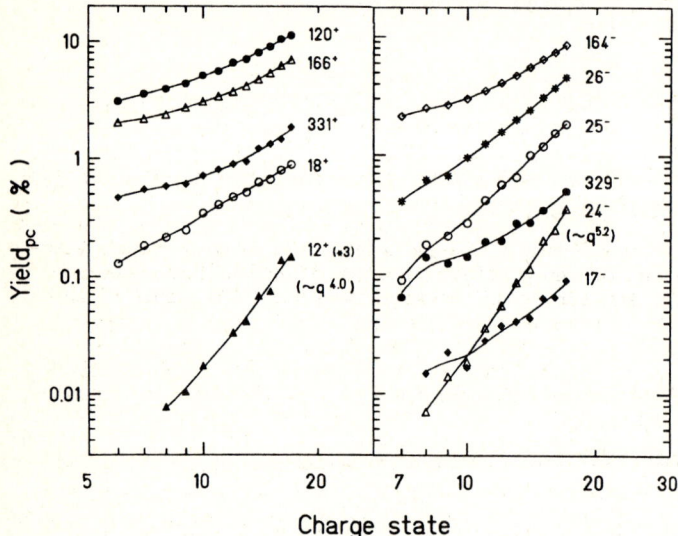

Fig.1 Yields of various secondary ions ejected from Phe by a Ca beam as functions of projectile-charge state

The two groups of ions - molecule specific ions and low mass fragment ions - show a second remarkable difference illustrated in Fig.2a: The figure presents yield curves of C^+ and $(M+1)^+$ ions measured with seven different beams. All C^+ yields fall more or less on the same curve being linear in log-log scale; they depend only on q. The yields of the parent ion $(M+1)^+$, however, are localized on separated curves differentiated by the kind of projectile. Whereas the later observation is explainable by the mentioned charge changing processes in layers underneath the surface, the q-dependence of C^+ is obviously not masked by charge changing. That means C^+ and other similar low mass fragment ions are ejected from the surface prior to energy dissipation from deeper layers. Their emission occurs probably in less than 10^{-12}s after the impact. They can be considered, therefore, as spectators of a relatively early stage of the desorption process. We suggest that these unspecific low mass fragments are generated close to the projectile's path in a zone of high excitation (additional comments see Ref.[5]).

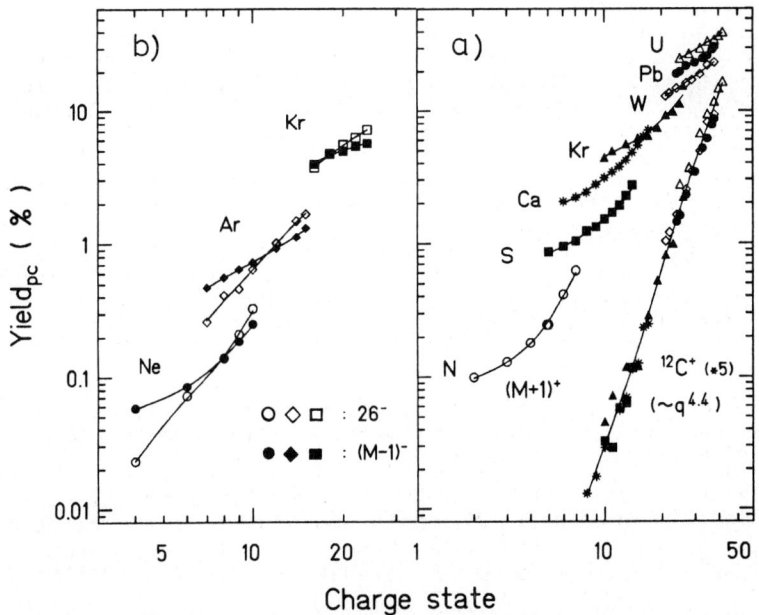

Fig.2 A comparison of secondary ion yields measured as function of charge state with various beams. The curves of the positive ions $(M+1)^+$ and C^+ have been obtained in Darmstadt (Fig.a), the curves of the negative ions $(M-1)^-$ and 26^- in Orsay (Fig.b)

In the case of negative ions (see Fig.2b) the different behaviour of the two groups is less pronounced. Also the yield curves of low mass ions are separated. Yields of the $(M-1)^-$ ion obtained with beams of the same q but different atomic number differ considerably, indicating a strong influence of charge changing underneath the surface.

When the charge state distribution of the incident beam is the same as inside the solid, an influence of charge changing should not be detectable. Therefore, yields measured with beams in the charge state $q = q_{eq}$ represent the true q-dependence - unless the excitation of ions penetrating the solid causes additional effects. Tombrello and co-workers point to this problem in connection with the forward-backward asymmetry of desorption from thin targets [1]. In the following we neglect ion excitation effects. Approximate values of q_{eq} were calculated with help of Ziegler's's formular for the effective charge Z_{eff} [6]. The yields Y(q) of one curve were then normalized to $Y(q_{eq})$ and plotted as function of $\Delta q = (q-q_{eq})/q_{eq}$. As illustrated in Fig.3 it turned out that all $(M+1)^+$ yields of Phe measured so far are describable by

$$Y(q)/Y(q_{eq}) = \exp(\lambda \cdot \Delta q) \qquad \text{with } \lambda = (15\pm 1)/q_{eq}$$

Exceptions are two data points obtained with N^{2+} and N^{3+}. The most interesting fact is the relation $\lambda \sim q_{eq}^{-1}$. It implies that charge

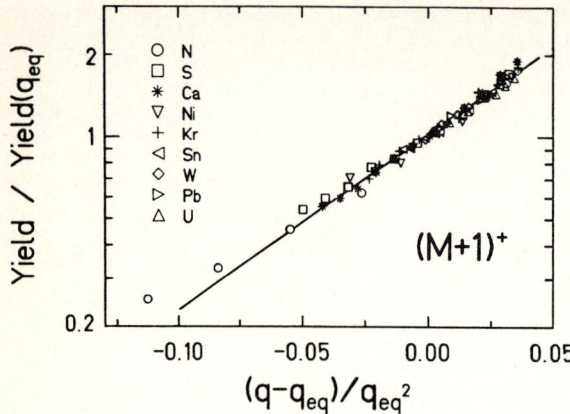

Fig.3 Normalized yields of the $(M+1)^+$ ion as function of the relative charge $q = (q-q_{eq})/q_{eq}$ divided by q_{eq}

changing increases towards larger q_{eq} or - what is probably the more appropriate quantity - towards a higher energy loss. We have two tentative explanations for this effect:

1. A higher energy density in the surrounding of the ions track enhances the energy transport from deeper layers to the surface. That means that heat conduction is increased.

2. The zone of high destruction close to the ions track is enlarged when the energy deposit increases. Molecule specific ions are ejected from areas further out. The direct influence of the primary charge state is reduced relative to long distance interactions from deeper layers.

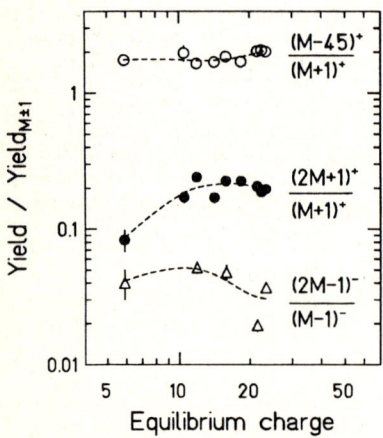

Fig.4 Relative yields of molecule specific ions as function of the mean equilibrium charge q_{eq} inside the solid

In order to exhibit the trend of $Y(q_{eq})$ over the whole range of projectiles, in Fig.5 the yields of the parent ions $(M\pm1)^\pm$ are plotted as function of q_{eq}. The data points have relatively large systematic errors, because they were obtained in three different accelerator runs. The data measured in Orsay and assigned by arrows were taken at a projectile velocity of 1.54 cm/ns. In Fig.5 they have been multiplied by a factor 1.4. As shown in Fig.5a, in log-log scale both yields can be approximated by linear curves. The $(M+1)^+$ yield follows a q_{eq}^3 depen-

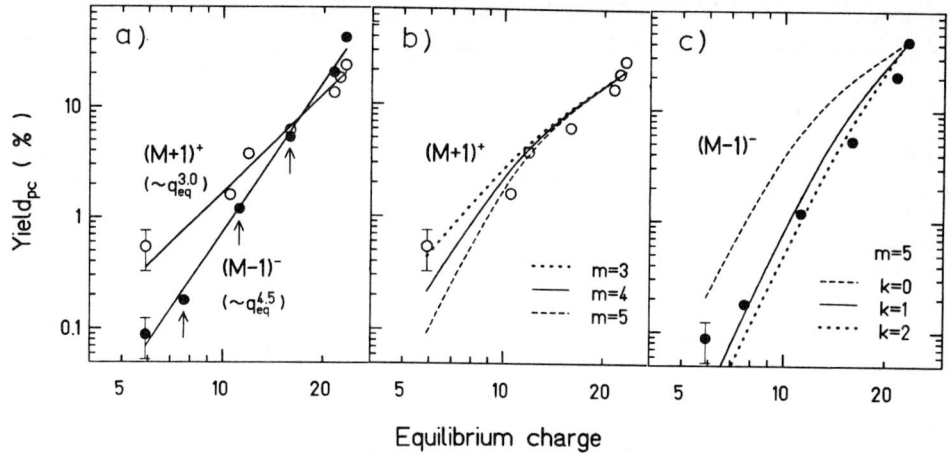

Fig. 5 Yields of the parent ions $(M\pm 1)^{\pm}$ as function of the equilibrium charge inside the solid. a) tentative q_{eq}-dependence of the yields, b) comparison of the $(M+1)^+$ yield curve with predictions of the ion track model [8], c) comparison of the $(M-1)^-$ yield curve with predictions of a modified ion track model [9]. Theoretical curves were normalized to one data point.

dence and the $(M-1)^-$ yield a $q_{eq}^{4.5}$ dependence. Nees et al [7] found a q^2 dependence for the positive parent ion of valine, but one should notice that they measured the yields as function of the equilibrium charge outside of the target.

In Fig.5b and 5a yields of the parent ion are compared with predictions of the ion track model of Hedin et al [8]. The intention of this model is to trace back desorption of large molecules to hits of electrons passing the molecules after the heavy ions impact. Depending on the molecular size, desorption is induced when the number of hits exceeds a minimum value m. The number of electrons per molecule is derived from the excitation-energy density in the vicinity of the ions track. With m = 4 per Phe molecule the $(M+1)^+$ yields are quite well described. In case of $(M-1)^-$ the trend of the data points could not be reproduced with any choice of m, but it needed only a minor modification of the model to improve the agreement: The radius of the inner zone of high damage, from where no undestroyed molecules are released, was increased as $k \cdot (dE/dx)^{1/3}$ (further details see Ref.[9]). That indicates in the framework of the ion track model that molecule specific negative ions are emitted from zones further out of the ions track than positive ions.

References

1 C.K.Meins, J.E.Griffith, Y.Qiu, M.H.Mendenhall, L.E.Seiberling, and T.A.Tombrello: Rad.Eff. 71, 13 (1983)
2 P.Håkansson, E.Jayasinghe, A.Johansson, I.Kamensky, and B.Sundqvist: Phys.Rev.Lett. 47, 17 (1981)
3 E.Nieschler, B.Nees, N.Bischof, H.Fröhlich, W.Tiereth, and H.Voit: Preprint, Inst.of Nucl.Phys., Univ. Erlangen 1984, FRG
4 W.Guthier, O.Becker, W.Knippelberg, A.Weikert, K.Wien, S.Della Negra, Y.Le Beyec, P.Wieser, and R.Wurster: Int.J.Mass Spectrom. Ion Phys. 53, 185 (1983)

5   K.Wien, O.Becker, and W.Guthier: to be published in Rad.Eff. 1985
6   J.F.Ziegler: Hand book Vok.5 of <u>The Stopping and Ranges of Ions in Matter</u>, ed. by J.F. Ziegler, Pergamon Press 1980
7   B.Nees, E.Nieschler, N.Bischof, P.Dück, H.Fröhlich, W.Tiereth, and H.Voit: Rad.Eff. <u>77</u>, 89 (1983)
8   A.Hedin, P.Håkansson, B.Sundqvist, and R.E. Johnson: Phys.Rev.B <u>31</u>, 1780 (1985)
9   S.Della Negra, Y.Le Beyec, O.Becker, and K.Wien: to be published

Particle Desorption from Non-Metallic Surfaces by High Energy Heavy Ions*

W. Guthier

Institut für Kernphysik, Technische Hochschule Darmstadt,
D-6100 Darmstadt, F.R.G.

1 Introduction

During recent years a growing number of experiments have focussed on particle desorption from insulating material induced by high energetic heavy ions in the MeV/u regime. In particular, the dependence of this energy sputtering on the energy and the charge state of the incident ions was studied [1-3]. It turned out that the sputtering yields are strongly related to the electronic energy loss. As will be shown in this work, a small portion of the sputtered particles is charged. These secondary ions are more accessible than the large neutral fraction. Sophisticated time-of-flight techniques were used to explore their properties in detail. That means, our knowledge about secondary ion emission is generally better than that about total particle desorption. On the other hand, most theoretical work about ion induced desorption deals with total sputtering. Ionisation is not taken into account and data of secondary ion emission are usually compared with desorption modelds, under the assumption that ion yields are proportional to sputter yields. The intention of this work is to prove this assumption by measuring ions and total yields from the same samples as function of the energy of the incident heavy ions.

One of the first experiments on sputtering from insulating material was that of W.L. BROWN et al [4]. They observed an enhanced sputtering from frozen gases irradiated by low-Z ions in the MeV region. One expects larger effects with high-Z ions which deposit an energy in the order of keV/Å into the electronic system of the target material. For instance, the energy loss of 1.4 MeV/u - ^{238}U in CsI is ~2keV/Å. There exist several proposals, how this intensive excitation of the electronic system is transferred into atomic motion [5-7]. Recent experiments by C.K. MEINS et al [8] with MeV/u-chlorine ions on UF_4 have shown that also in this case the sputter yield is enhanced.

In the present experiments the energy of the incident ions is varied in the range of 0.4 to 1.4 MeV/u. Cesium iodide and europium oxide samples are irradiated by krypton, xenon and uranium ions. All results agree in the fact that the total desorption yield is three or four orders of magnitude higher than that of the charged particles.

2 Experimental techniques

Cesium iodide and europium oxide were evaporated onto Al-foils under a vacuum pressure of 10^{-6} torr. The evaporation of the europium oxide was performed with help of an electron beam and that of CsI by thermal heating. CsI was evaporated under two different conditions: First, the

---

* Supported by the Deutsche Forschungsgemeinschaft; part of Thesis

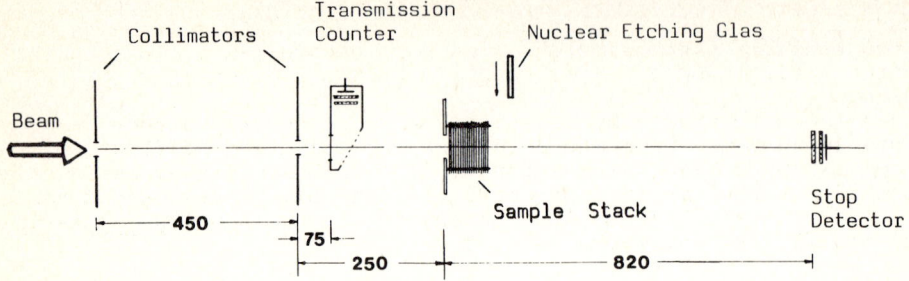

Fig.1 Schematic drawing of the experimental arrangement. The distances are given in millimeters (dimensions not in scale)

CsI molecules were released from the liquid phase in the melting pot, and secondly, the CsI molecules sublimated during a very slowly increasing temperature. In the latter case only part of the material melted. The thickness of all samples was 100 µg/cm^2.

The experimental arrangement is given schematically in Fig.1. In front of the target a carbon catcher is mounted within a distance of two millimeters. The carbon foils (100 µg/cm^2) were produced by the target laboratory of the GSI. All the sputtered particles should be caught on the assumption that the sticking probability is one. Additional foils allow to degrade the projectile energy. Such triplets of target-catcher-degrader foils build up an experimental stack for one beam exposition. Generally, two basic measurements were made: On the one side, the incident beam penetrates first the Al-carrier foil and then the sample -this is called "behind"-, and on the other side, the beam penetrates first the sample and then the Al foil -this is called "head". A pair of collimators reduces the beam width to a diameter of 6 mm. Concerning the catcher diameter of 15 mm, all particles ejected from the surface should be collected independently from the increasing angular spread of the beam passing more and more foils. A transmission counter in front of the stack registers the beam projectiles in order to determine the total number of incident primary ions. The rate of primary ions was restricted to not more than $3 \cdot 10^4$ s^{-1}. With a total number of 10^9 penetrating heavy ions it was guaranted that at the sample surface an area with a diameter of at least 1000 Å^2 was hit only once [9].

During the time of exposition there was an additional target-catcher arrangement in the vacuum chamber. It was used to determine the rate of sublimation of sample material on identical conditions. A stop detector at the end of the beam line allowed to measure a time-of-flight spectrum of the beam particles in order to determine their velocity. A nuclear etching glas could be moved into the beam. It was used to estimate the number of incident projectiles and to prove the adjustment of the beam spot on the targets. The target stacks were irradiated at a parasitic beam of the UNILAC accelerator at the GSI/Darmstadt (maximum energy 1.4 MeV/u).

The amount of sputtered material was determined by neutron activation analysis (NAA). For this purpose, after irradiation the carbon catcher foils were packed into polyurethane boxes and exposed to $2 \cdot 10^{12}$ thermal neutrons cm^{-2} s^{-1} in a nuclear reactor at Heidelberg[+]. After activation

---

[+] The activation by neutrons was performed at the Deutsche Krebsforschungszentrum TRIGA-HD II research reactor at Heidelberg.

the catcher foils were removed from the boxes. Their γ-ray spectra were analysed by means of Ge(Li)-detectors at the GSI. With help of several activated standard samples of the same sample material the procedure of neutron activation and γ-ray measurement was calibrated.

With targets prepared by the same method as described above, secondary ion mass spectra were measured by means of a time-of-flight method described elsewhere [10]. The secondary ion yields measured as function of projectile velocity were corrected for dead time losses and detection efficiency. In the following figures errors due to counting statistics and neutron flux are given.

3 Experimental results

The first results obtained with krypton and thorium on CsI are presented in Fig.2. The data points scatter considerably, a significant difference between krypton and thorium irradiation is not observed. In case of thorium, head-on irradiation seems to generate a higher yield than from behind.

A reason for the ambiguous results could be that structure and coverage of the sample surfaces were different. As seen in Fig.3 the carrier foil is covered by crystals of various sizes. Under air the structure of the

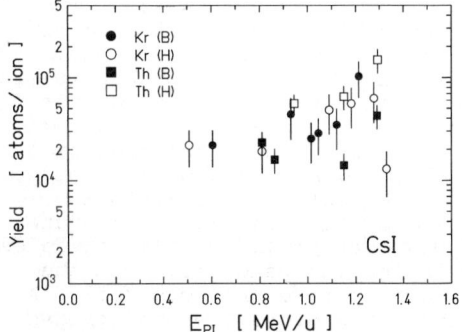

Fig. 2 Yield of Cs sputtered from CsI by ^{84}Kr and ^{232}Th ions as function of projectile energy

Fig.3 Target surface of CsI observed by means of electron microscopy - REM - (0.8 cm ≙ 10^4 Å). The CsI film was produced by the second evaporation method. Sample surfaces are shown before (left) and after (right) irradiation with 1.4 MeV/u uranium ions.

sample surface changes. Sometimes even at a sample thickness of 2400 Å, the small portions of the carrier foil are visible through the CsI layer. Another problem is sublimation of the sample during the irradiation period. Experimentally it was found that the sublimation rate is 10^9 molecules per hour, causing considerable background at long exposition times. The samples corresponding to the results in Fig.2 were prepared by evaporation out of the melted material.

Using the second evaporation method, the total cesium yield was found to be one order of magnitude lower than in the case of the first method (see Fig.2). The yields presented in Fig.4 were measured with improved techniques. Samples were irradiated by $2.0 \cdot 10^4$ U ions s^{-1}. This time sublimation was at least a factor ten lower than in the first case, indicating that sublimation rate and sputtering yield are eventually correlated.

As seen in Fig.4 the yield of the Cs^+ ion is about 10^3 times smaller than the total Cs yield. It is interesting that the total yield is with in the error bars proportional to $(dE/dx)^2$, whereas the Cs^+ yield decreases towards higher projectile energies.

Europium oxide has two advantages compared to CsI: Sublimation is negligible and the neutron activation cross section is 135 times higher.

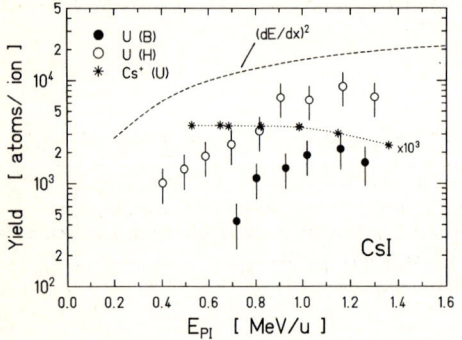

Fig.4 Yield of Cs and Cs^+ sputtered from CsI by ^{238}U; head (o) and behind (●) bombardment. The dotted line is the Cs^+ secondary ion yield obtained with ^{238}U ions.

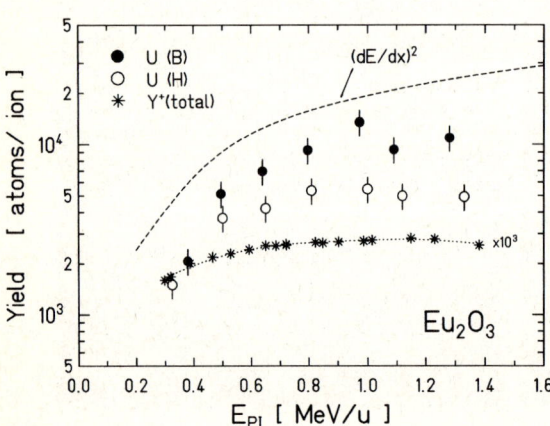

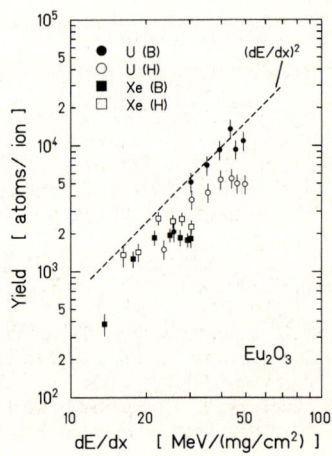

Fig. 5 Yield of Eu sputtered from europium oxide by ^{238}U. The dotted line is the sum of all positive secondary ions containig Eu atoms.

Fig. 6 Yield of Eu as function of electronic stopping power.

As shown in Fig.5 the total yield curves of europium follow again more or less $(dE/dx)^2$. Above 0.5 MeV/u the slope of the ion yield curve is similar to that of the total yield. The sputter yields of europium oxide have been investigated with U and Xe beams. In Fig.6 corresponding yields are plotted as function of dE/dx. The trend of the data points follows $(dE/dx)^2$, but in the high energy part of each data set certain deviations appear. Similar deviations are known from secondary ion yield curves.

Quite remarkable are the differences between yields measured with head on irradiation and irradiation from behind. Both types of data were obtained with the same stack of targets. In two cases (U→CsI, Xe→Eu_2O_3) the head on yields were higher, in one case (U→ Eu_2O_3) they were lower than the yields from behind. So far, the situation is not clear. More examples are in preparation.

4 Conclusion

As shown in the preceding chapter the yield of secondary ions emitted from CsI and Eu_2O_3 by heavy ion irradiation is three or four orders of magnitude smaller than the total desorption yield. A similar behaviour was observed with organic samples like oxophenylarsine, but in this case sublimation during the irradiation period causes severe difficulties. A relation to $(dE/dx)^2$ was only tentatively observed; deviations are similar for ion and total desorption.

Acknowledgments

The author wishes to thank Prof. Dr. K. Wien for many helpful discussions and also Dr. Maier-Borst and D. Jünger from the Deutsche Krebsforschungszentrum Heidelberg for providing the research reactor. Further thank to H. Folger, Dr. G. Wirth, L. Dörr, H. Penninger and G. Buggisch for their helpful cooperation during the GSI experiments. Also the assistance of the computer staff and the excellent work of the UNILAC operators of the GSI are gratefully acknowledged.

References

1. P.Dück, H.Fröhlich, N.Bischof, H.Voit, Nucl. Instr. and Meth. 198 39 (1982)
2. P.Håkansson, I.Kamensky, B.Sundqvist, Nucl. Instr. and Meth. 198 43 (1982)
3. W.Guthier, O.Becker, S.Della Negra, W. Knippelberg, Y.Le Beyec, U.Weikert, K.Wien, P.Wieser, R.Wurster, Int. J. Mass. Spectrom. Ion Phys. 53 185 (1983)
4. W.L. Brown, W.M.Augustyniak, E.Brody, B.Cooper, L.J.Lanzerotti, A.Ramirez, R.Evatt, R.E.Johnson, Nuc. Instr. and Meth. 170 321 (1980)
5. G.H.Vineyard, Rad. Eff. 29 245 (1976)
6. L.E.Seiberling, J.E.Griffith, Y.Qui, T.A.Tombrello, Rad. Eff. 52 201 (1980)
7. C.G.Watson, T.A.Tombrello, Rad. Eff. 89 263 (1985)
8. C.K.Meins, J.E.Griffith, T.A.Tombrello, M.H.Mendenhall, L.E.Seiberling, Rad. Eff. 71 13 (1983)
9. C.Riedel, R.Spohr, Rad. Eff. 42 69 (1979)
10. W. Guthier, Thesis TH Darmstadt, in preparation

^{252}Cf-PDMS: Multiplicity of Desorbed Ions and Correlation Effects

L. Schmidt and H. Jungclas

Kernchemie, Fachbereich Physikalische Chemie, Philipps-Universität,
D-3550 Marburg, F.R.G.

1. Introduction

The basic principles of high energy heavy-ion induced desorption are summarized elsewhere [1,2] and assumed to be known. Secondary ions are desorbed either by the impact of a fission fragment from a Cf-252 source [1] or of a heavy ion beam pulse, e.g. I-127 [3] and their time-of-flight is analyzed event by event.

In the 1970s, little attention had been paid to the possibility to register several secondary ions (stop events) for each ionization event. A first correlation investigation had been performed by means of a common time to analog converter (TAC), detecting single stop events in a coincidence mode [4]. The use of a multistop time to digital converter (TDC) gave the chance to count the number n of detected secondary ions [3,5] and to investigate correlation effects [5]. The progress made by the Orsay group became possible by a multistop TDC which employs a buffer for 31 stop events per interrupt (n ≤ 31) [6]. We utilized a similar TDC module for our measurements.

2. Multiplicity of Desorbed Ions

From each ionization event (here called interrupt i or start event for the time-of-flight measurement) secondary ions are desorbed with the multiplicity m_i. Due to the efficiency and the time resolution of the detection system just the number n_i of ions are detected. These generate n_i stop pulses (per interrupt), each resulting in an increment at the appropriate time-of-flight t in a spectrum with T time bins. At the end of the acquisition period (last interrupt i=I) the quantity $M = \sum_i m_i$ ions are desorbed, however, $N = \sum_i n_i$ ions are detected. This leads to an overall efficiency $\varepsilon = N/M$. The multiplicity distribution of desorbed ions and its mean value $\overline{m} = M/I$ contain interesting information on the ionization and desorption process. Since the true multiplicity can not be measured, we investigate and discuss the number of detected ions with its mean value $\overline{n} = N/I$.

Using a sample of Verapamil (mol.wt. 454) [7], which has been prepared by evaporation, we have analyzed the rate R_{tot} of interrupts, where the start event is related to n stop events. In Fig.1 R_{tot} is plotted vs n. The integral total rate $\sum_n R_{tot}(n) = I$ equals the number of analyzed start events (interrupts I); the mean value is $\overline{n_{tot}} = 2.8$.

Next, a time window has been set on the molecular ion peak by software. This allows to analyze those ions which are desorbed and detected in coincidence (in

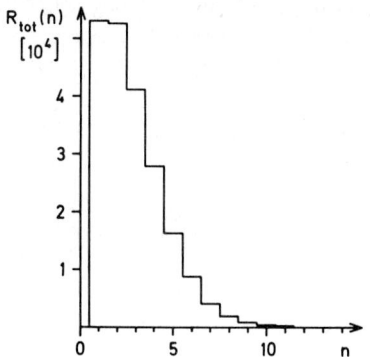
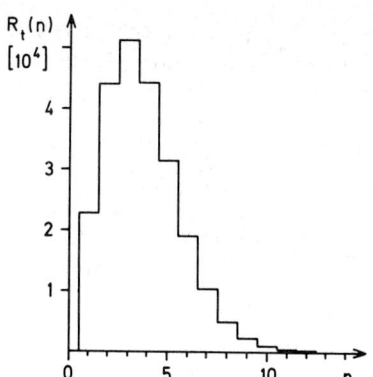

Fig.1 Measured n-distribution, i.e. the total rate of start events with n detected ions using a Verapamil sample

Fig.2 Same distribution as in figure 1 but with the condition, that 1 of n detected ions is a molecular ion

the same interrupt) with a molecular ion at time-of-flight t. The rate of interrupts, where 1 of n stop events fulfills this coincidence condition t, is shown in Fig.2. As not all start events (from the quantity I) are related with a detected molecular ion, the total rate in this second measurement is smaller than in the first case: $\sum_n R_t(n) = I_{eff} < I$. However, the mean value is larger: $\overline{n_t} = 3.7$. This observation is true for any coincidence condition t, as demonstrated before [3,5] but misunderstood [3,1], as we will discuss later. First we clarify the statistical background by discussing the probability of multiple ion detection.

Assume there are n random stop events per start, thus the resulting time of flight spectrum (in T channels) shows random noise. Then the probability $P_t(n)$, that 1 of n stop events falls into a certain time bin t, will increase with the number n till it reaches at n=T the saturation: $P_t(n<T) = n/T$, $P_t(n \geqslant T) = 1$. From the experimental data above one can calculate the probability to detect (n-1) ions in coincidence to a preset ion in a time window t: $P_t(n) = R_t(n) / R_{tot}(n)$. As both rates are based on two different data sets, the rates had to be normalized on the quantity I. The result is plotted in Fig.3 for three different coincidence windows. First, $P \propto n$ for $n \leqslant 4$, as we have shown above for a random noise spectrum. Second, the slope of this linear section increases with the count rate in each coincidence window. Third, a maximum is achieved for $n \approx 10$ and finally P even descends with increasing n. This observation is more obvious in Fig.4, which is based on measurements with reduced detection efficiency (by setting a higher discriminator level only larger stop pulses are accepted).

In order to understand this behaviour (deviation from the linear relation $P \propto n$), we measured this probability for an arbitrary time window, just containing noise. The result (Fig.5) reveals that the noise increases with a higher order of n.

As the multiplicity m of desorbed ions can not be measured directly, the $R_t(n)$-distribution (Fig.2) has been unfolded according to the dead time $\Delta t = 100$ns of the TDC (after each stop event) and an estimated detection efficiency $\varepsilon_d = 0.55$. The calculated $R_t(m)$-distribution (Fig.5) for (m-1) ions, which are desorbed coincidentally with a molecular ion, shows a mean multiplicity $\overline{m}_t = 6.5$.

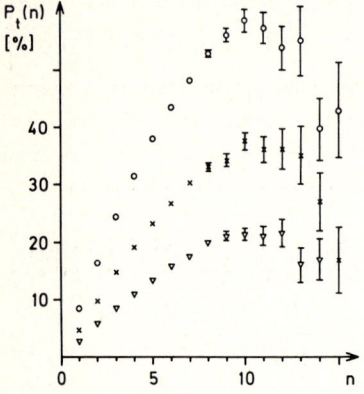

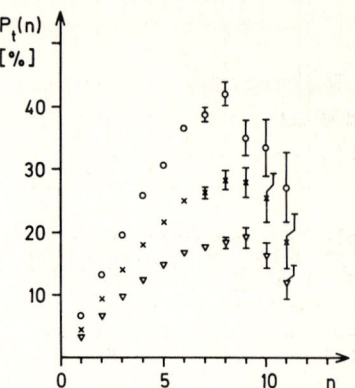

Fig.3 Detection probability of 3 selected ions in interrupts with n stop events from a Verapamil sample (o m/z 454±1, x m/z 165, ▽ m/z 44)

Fig.4 Same as figure 3 but at a reduced detection efficiency

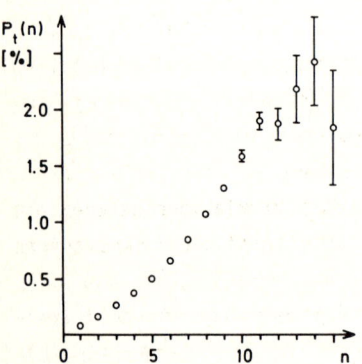

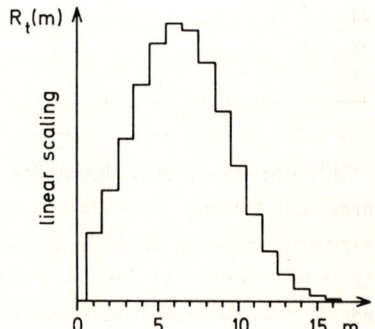

Fig.5 Probability of noise events (preset time window) in interrupts with n different stop events

Fig.6 Theoretical multiplicity distribution with the condition that 1 of m desorbed ions is a molecular ion (unfolded from figure 2)

3: Correlation Effects

When two ion species A and B are related in the ionization process, it should be possible to prove a correlation between the two corresponding peaks. Interrupts with large number n could wash out a correlation effect due to random coincidences and due to the raised noise. Thus we did not analyze the complete data set, but restricted ourselves to interrupts with n=2.

We investigated the correlation between pairs of peaks in the spectrum of Verapamil. We measured the conditional probabilities $c = P_n(B/A)$ i.e. for B when

A is true (coincidence) and $a = P_{n-1}(B/\overline{A})$ i.e. for B when A is false (anticoincidence). A correlation coefficient r is defined:

correlation   (c > a): $r = 1 - a/c$ $(0 \leq r \leq +1)$
anticorrelation (c < a): $r = c/a - 1$ $(-1 \leq r \leq 0)$
independence   (c = a): $r = 0$.

The results are summarized in Table 1, showing how the probability to produce and detect an ion B depends on the condition whether or not an ion A is produced during an ionization event.

Table 1 Correlation coefficients (%) for pairs of peaks in the positive ion spectrum of Verapamil. The condition A, i.e. the peak on which a time window has been set, is listed in the first line.

m/z A \ B	454	303	165	151	44	1
454	/	23 ±1	13 ±1	8 ±1	11 ±1	8 ±1
303	21 ±1	/	6 ±2	5 ±1	-2 ±2	3 ±2
165	14 ±1	11 ±2	/	17 ±1	7 ±2	5 ±2
151	2 ±2	4 ±3	16 ±3	/	8 ±3	1 ±3
70	-4 ±3	0 ±4	8 ±4	9 ±2	14 ±3	-3 ±4
44	2 ±2	4 ±3	12 ±2	5 ±2	/	10 ±2
39	-2 ±3	3 ±4	4 ±4	-4 ±3	16 ±3	15 ±3
1	0 ±2	3 ±2	-2 ±2	-3 ±2	5 ±2	/

There is no save evidence for an anticorrelation, but correlations between the three main peaks, i.e. the quasimolecular ions (m/z 454±1) and the fragment ions (m/z 303, m/z 165) and for some other ion combinations are significant (r > 0.1). The true correlation will be even stronger, as the coefficients should be corrected for the detection efficiency.

4. Discussion

The obvious difference between the total rate of detected ions (Fig.1) and the rate of coincidentally desorbed (e.g. with a molecular ion) and detected ions (Fig.2) has been stressed before [3,5]. But the suggested interpretation that molecular ions are preferentially formed by "super tracks" [1] is false, as it does not take into account a simple fact: (1) The probability to detect a molecular or any other ion increases with the multiplicity m of desorbed ions, resp. the number n of detected ions. When the probability relation $P_+(n)$ is known, the conditional rate (e.g. Fig.2) can even be calculated from the total rate (Fig.1).

Molecular ions, just as other ions, are detected at any number n with a constant relative intensity (e.g. Fig.3). However, at higher n the absolute ion intensities decrease, as the competition by noise increases. Two conclusions arise:
(2) The signal-to-noise ratio descends with an increasing number n of detected events per interrupt. (3) A restriction to small n values (4) leads to cleaner spectra.

Two facts can establish the wrong impression that the mean multiplicity of desorbed ions is very small. First, single fission fragments are used to ionize sample molecules. Second, the time-of-flight of secondary ions is acquired event by event. On the other hand, it is known that a fission fragment can sputter up to 10^4 molecules as neutral clusters [8] and ionizes approximately 1% of these. This agrees with our result on the unfolded multiplicity distribution (Fig.5). Our unfolding procedure [9] does consider the detection efficiency ε_d and the dead time Δt of the TDC. However, it does not take into account that coincident ions with same m/z contribute to the same stop signal. In order to estimate the number of coincident ions, one would need a pulse height analysis of the stop signals. For molecules like Verapamil with a simple fragmentation pattern, the mean multiplicity might be larger than the number of different fragment ions. This effect of multiple ions per stop signal, which should cause an amplitude dispersion of stop signals for the same m/z, would be severe. This implies two further conclusions: (4) The "true" multiplicity of desorbed ions can be even higher than shown in Fig.6 ; the number of secondary ions produced by a single fission fragment might reach the order of 10^2. (5) The chance to detect a stop signal at a certain m/z increases with the number of coincidentally stopping ions.

The results of our correlation measurements are preliminary. We have shown that correlations between secondary ions in Cf-252-PDMS can be analyzed and seem to be strong. There are not yet enough data for evaluation and interpretation of the ionization process. In further measurements more complex molecules, positive and negative ions and the influence of sample thickness should be investigated

References

1 R.D. Macfarlane: Anal. Chem. 55, 1247 (1983)
2 R.D. Macfarlane: Proc. 2nd Int. Workshop Ion Formation Org. Solids, Münster 1982, Springer Ser. Chem. Phys. 25, 32 (1983)
3 B. Sundqvist, P. Håkansson, J. Kammensky and J. Kjellberg: same location as ref. 2, 52 (1983)
4 N. Fürstenau, W. Knippelberg, F.R. Krueger, G. Weiss and K. Wien: Z. Naturforsch. 32a, 711 (1977)
5 S. Della Negra, D. Jacquet, J. Lorthiois, Y. Le Beyec, O. Becker and K. Wien: Int.J.Mass Spectrom. Ion Phys. 53, 215 (1983)
6 E. Festa and R. Sellem: Nucl. Instr. Meth. 188, 99 (1981)
7 H. Jungclas, H. Danigel and L. Schmidt: Org. Mass Spectrom. 17, 86 (1982)
8 J.P. Biersack, D. Fink and P.J. Mertens: J. Nucl. Mat. 53, 194 (1974)
9 E.A. Koop: unpublished thesis, Universität Marburg (1985)

Part II

Secondary Ion Mass Spectrometry (SIMS)

Surface Organic Reactions Induced by Ion Bombardment

R.G. Cooks[1], *B.-H. Hsu*[1], *W.B. Emary*[1], *and W.K. Fife*[2]

[1] Department of Chemistry, Purdue University, West Lafayette, IN 47907, USA
[2] Department of Chemistry, Indiana University-Purdue, University at Indianapolis, Indianapolis, IN 46223, USA

1. Introduction

While chemical reactions, for example cationization, are common in desorption ionization methods, complex bonding processes are not [1-3]. Electron or ion transfer, clustering and fragmentation (often by simple gas phase dissociation) are often involved in ion formation when organic molecules are subjected to particle impact. Inorganic complexes and organometallic compounds often undergo more complex processes, including ligand exchange and other condensed phase behavior [4-5]. MICHL [6] has shown that simple nitrogen oxides in frozen matrices are very reactive upon argon ion bombardment, and several authors have demonstrated that chemical reduction can occur in glycerol under fast atom bombardment conditions [7-11].

This paper is concerned with chemical reactions and strong matrix interactions in samples bombarded by keV energy particles under static and dynamic secondary ion mass spectrometry (SIMS) conditions using solid and liquid matrices [12]. The chemical reactions induced by ion beams are of intrinsic and analytical interest, while the matrix interactions have implications for the mechanism of static SIMS as well as its analytical performance [13].

2. Beam-Induced Chemical Reactions with Solid Matrixes (Static Conditions)

The SIMS spectrum, taken under conventional static conditions, of benzotriazole (BTA) supported on silver contains not suprisingly a base peak at m/z 120, $(M+H)^+$. Also present are $M^{+\cdot}$ at m/z 119 (53% relative abundance R.A.) and an anomalous peak at 133 with 92% R.A.. Isotopic labeling (^{15}N at the 2 position) established that $(M+14)^+$ corresponds to $(M+N)^+$, a most unexpected result. Because intermolecular N transfer is involved, the overall reaction for the protonated form of the dimer can be written as:

$$(BTA)_2H^+ \longrightarrow \text{[benzotriazole-NH]}^+ + \text{[benzene-di-NH]}$$

The corresponding intermolecular CH transfer process occurs in the imidazole analog, although transfer of a sulfur atom is not observed in benzothiazole. The nitrogen transfer reaction occurs also in the 5(ring)-methylated analog but not in the N-methyl compound, perhaps because the bulky methyl group sterically hinders the formation of a tightly bound dimer which may be necessary for reaction to occur. The N-methyl compound shows peaks at $(M+15)^+$ and $(M+14)^{+\cdot}$, but the latter is believed to be due to fragmentation of the former, the intermolecular transmethylation product.

Further information on nitrogen atom transfer in BTA was sought through the study of matrix effects, results for which are accumulated in Table I.

Table I: Matrix effects in static SIMS

Matrix	Compound	Substrate	Matrix/Comp'd Ratio	Molecular Ions, m/z	Fragment Ions, m/z	Matrix Effect
$MgCl_2$	(benzotriazole with H_3C)	Ag	2	$134(M+H)^+$ $133(M^{+\cdot})$	104, 105	Suppresses Analyte Ions Except Beam-Induced Reaction Product, $(M+N)^+$
NH_4Cl	"	Ag	2	$134(M+H)^+$ $133(M^{+\cdot})$	104, 105	Reduces $(M+H)^+$ Ions, and Enhances $(M^{+\cdot})$ Ions
$MgCl_2$	(benzotriazole with CH_3)	Ag	2	$133(M^{+\cdot})$	105	(i) Enhances m/z 147 Ions (ii) Reduces Self-Methylation Product, $(M+CH_3)^+$
NH_4Cl	"	Ag	2	$133(M^{+\cdot})$	105	Reduces Both Ions of m/z 147 and m/z 148
Ag Overlayer Deposition	(benzotriazole)	Cu	--	$119(M^{+\cdot})$ $120(M+H)^+$	91	Reduces $(M+H)^+$ and $M^{+\cdot}$ Ions Suppresses Beam-Induced Reaction Product, $(M+N)^+$
Cu Overlayer Deposition	"	Ag	--	$119(M^{+\cdot})$ $120(M+H)^+$	91	"

Some noteworthy features are: (i) The reaction does not occur in a glycerol matrix under high flux (FAB) conditions. (ii) Addition of NH_4Cl (known to suppress intermolecular alkylation in carnitine) [14] has relatively smaller effects on the BTA reaction, suggesting that a tightly bound dimer, perhaps I, is the reactant. (iii) p-Toluenesulfonic acid enhances $(M+H)^+$ at the expense of $(M+N)^+$, as would be expected for a basic compound in an acidic matrix. (iv) A silver matrix, generated by codeposition of BTA and sputtered silver, appears to be a successful method of isolating BTA molecules; the mass spectrum contains $(M+Ag)^+$, $(2M+Ag)^+$, and $(M-H+2Ag)^+$ ions but interestingly no $(M+H)^+$, $M^{+\cdot}$, or $(M+N)^+$ species. (v) Deposition of a silver overlayer on BTA supported on copper suppresses $(M+N)^+$ and gives chiefly $(M+H)^+$, $(M+Ag)^+$, and $(M-H+2Ag)^+$. (vi) $MgCl_2$, a matrix with pronounced effects, was found to suppress analyte peaks except that due to $(M+N)^+$. Other data (presented below) indicate that one possible effect of $MgCl_2$ is to make the generation of free gas phase ions more difficult. This is consistent with the observation regarding the relative enhancement of the ion $(M+N)^+$, a product of intermolecular reaction.

I

At this point it is well to recall that matrix effects in desorption ionization may be of three types: (a) effects on the types of ions generated, (b) effects on total ion yields and (c) effects on relative ion abundances [15]. A solid matrix studied extensively in this laboratory, NH_4Cl, can be used to illustrate this behavior. The salt enhances total ion yields and there is a marked decrease in the extent of fragmentation [15]. The latter behavior has been explained as the result of desolvation of a cluster species with a corresponding reduction in internal energy of the organic ion. One can describe ammonium chloride as softening the ionization conditions. On the other hand, magnesium chloride often has the opposite behavior, leading to increased fragmentation, and this matrix can be said to harden the ionization conditions.

Figure 1 illustrates the behavior of $MgCl_2$ in two separate cases and Table II contains a summary of a number of others.

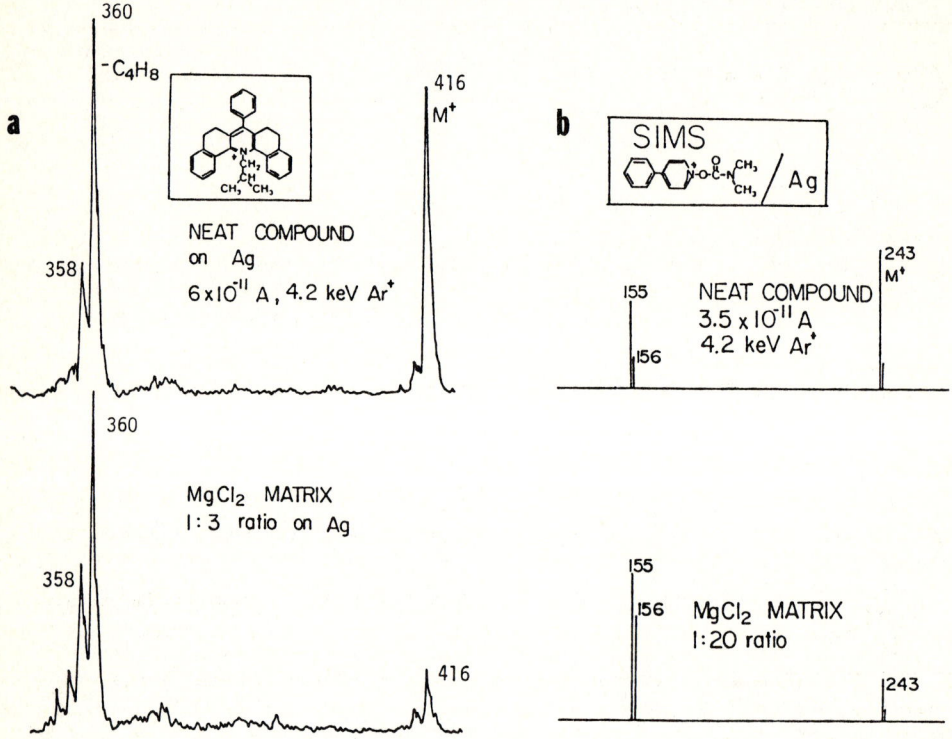

Figure 1. Matrix effect of $MgCl_2$ on the SIMS spectra of pyridinium salts

The positive ion spectra of 1-isobutylpyridinium perchlorate with, and without, magnesium chloride matrix are shown in Fig. 1a. The intact cation of the neat compound (m/z 416) is desorbed, and hydrogen rearrangement with elimination of neutral butene and butane result in ions at m/z 360 and 358, respectively. This is a typical SIMS spectrum of pyridinium salts, which usually consist of the intact cation and a few structurally informative fragment ions. The predominant fragment ion at m/z 360 is the base peak and the molecular ion at m/z 416 occurs with 82% relative abundance. In the presence of $MgCl_2$, which is physically mixed with the salt, the relative extent of fragmentation increases dramatically. The R.A. of the intact cation is reduced to about 20% while the fragment ion 360^+ is still the base peak in the spectrum. The Sims spectrum of 1-dimethylcarbonyl-oxo-4-phenyl pyridinium contains the intact cation as the base peak at m/z 243 (Fig. 1b). Dominant fragment ions occur at m/z 155 (60% R.A.) and m/z 156 (22% R.A.) which correspond to the radical cation of 4-phenylpyridine and the 4-phenylpyridinium cation, respectively.

Upon 20-fold dilution with magnesium chloride, the intensity ratio between the intact cation at m/z 243 and the fragment ions at m/z 155 and m/z 156 show a remarkable increase. This reflects an increased degree of fragmentation due to the addition of $MgCl_2$. 1-Benzoyloxypyridinium ion, which has an analogous structure to the alkyl 1-acyloxypyridinium, behaves quite differently in the presence of a $MgCl_2$ matrix, as evidenced in its SIMS spectrum (Table II).

Table II: Effects of $MgCl_2$ matrix*

Compound	Substrate	Matrix/Comp'd Ratio	Molecular Ions, m/z	Fragment Ions, m/z	Matrix Effect
[structure]	Ag	20	243(M^+)	155,156	Enhances Fragment Ions
[structure]	Ag	3	415(M^+)	359	Enhances Fragment Ions
[structure]	Ag	2	200(M^+)	105	Enhances Molecular Ions
$(CH_3CH_2)_4N^+$	Ag	50	130(M^+)	86,100	"
[structure]	Ag	2	137(M^+)	80	Almost No Effect
[structure]	Ag	7	137(M^+)	93,94	Suppresses Analyte Ions

* Analyte was physically admixed with $MgCl_2$ and spectra were taken under static SIMS conditions.

The intact cation of this salt can be observed at m/z 200 as expected. Cleavage of the oxygen benzoyl carbon bond results in the base peak ion at m/z 105, assigned as $C_6H_5CO^+$. However, when this salt was admixed with 2-fold excess of $MgCl_2$, a marked decrease in fragmentation was observed. Clearly the behavior of $MgCl_2$ is not simple and the possibility that several effects may be operating exists. Much remains to be learned of SIMS mechanisms through study of the chemical effects of matrix when analytes are subjected to an energetic beam of particles. It is possible that matrices which harden ionization conditions might be of value in characterizing biological molecules, which typically fragment poorly, by increasing the internal energy of the molecular ions when subjected to collisional activation.

3. Liquid Matrixes (Dynamic Flux Conditions)

Although this paper has presented data for one particular class of compounds using solid matrixes and particle desorption, liquid matrixes under high flux conditions also yield interesting chemical interactions with the analyte molecule. These reactions are unlikely to compromise the analytical utility of molecular SIMS. A case which illustrates both points is the formation of an ion $(SbCl_2 \cdot glycerol)^+$ when $SbCl_3$ in a glycerol matrix is examined under conventional fast atom bombardment conditions. Unexpectedly, it appears that this and many other ions undergo intramolecular isomerization under FAB conditions.

$$\begin{array}{c}Cl\\ \diagdown \\ Sb\cdots O \\ \diagup \\ Cl \end{array}\!\!\!\!\begin{array}{c}+\\ H\\ \diagup \\ \big) \\ OH\end{array} \rightarrow \begin{array}{c}ClH\\ \diagdown \\ Sb-O \\ \diagup \\ Cl \end{array}\!\!\!\!\begin{array}{c}+\\ \\ \big) \\ OH\end{array} \rightarrow \begin{array}{c}ClH\quad O\\ \diagdown \quad \diagup \\ +Sb \\ \diagup \quad \diagdown \\ ClH\quad O\end{array}\!\!\!\big) \qquad \Big(\!\!\begin{array}{c}OH\\ \diagup\\ \diagdown\\ OH\end{array} = glycerol$$

The rearrangements illustrated are demonstrated by the behavior of these ions upon collision-induced dissociation. The facile loss of HCl indicates that rearrangement to the HCl solvated structure has occurred. The daughter spectra of $(glucose \cdot G_n+H)^+$, n=1-2, G=glycerol, shows that dissociation of the diglycerated adduct yields predominantly a desolvated ion while $(glucose \cdot G+H)^+$ loses exclusively H_2O, indicative of a covalently bound ion. Other systems (organic and inorganic) show analogous behavior (refer to Table III).

Table III. Summary of fragmentation of organic and inorganic adducts*

Analyte	Parent Ion	Fragment lost (relative abundance)†
$SbCl_3$	$(SbCl_2G_2)^+$	HCl (100), G (32)
	$(SbCl_2G)^+$	HCL (100), G (5)
$FeSO_4$	$(FeSO_4 G_2+H)^+$	G (100)
	$(FeSO_4 G+H)^+$	SO_3 (100)
$MgSO_4$	$(MgSO_4 G_2+H)^+$	G (100)
	$(MgSO_4 G+H)^+$	SO_3 (100)
Glucose	$(Glucose\ G_2+H)^+$	G (100)
	$(Glcose\ G+H)^+$	H_2O (100)
Fructose	$(Fructose\ G_2+H)^+$	G (100)
	$(Fructose\ G+H)^+$	H_2O (100)

* Conditions: primary particle flux=$3.18 \times 10^{-5} A cm^{-2}$; Energy=7keV daughter spectra were obtained using B/E linked scans

† G=glycerol

4. Acknowledgement

This work was supported by the National Science Foundation (CHE-81-11425). B.-H. H. acknowledges a fellowship from the IBM corporation. We thank Dr. Robert J. Day for valuable suggestions.

5. References

1. Benninghoven A., Ion Formation from Organic Solids, Proceedings of the Second International Conference, Münster 1982, A. Benninghoven, ed., Springer-Verlag, Berlin-Heidelberg-New York, 1983.
2. Pachuta S.P., Cooks R.G., ACS Symposium Series 291 (1985), 1.
3. Busch K.L., Cooks R.G., Walton R.A., Wood K.V., Inorg. Chem. 23 (1984), 4093.

4. Macfarlane R.D., Ion Formation from Organic Solids, Proceedings of the Second International Conference, Münster 1982, A. Benninghoven, ed., Springer-Verlag, Berlin-Heidelberg-New York, 1983.
5. Miller J.M., Adv. Inorg. Chem. and Radiochem. $\underline{28}$ (1984), 1.
6. Michl J., Int. J. Mass Spectrom. Ion Phys. $\underline{53}$, 255 (1983).
7. Clayton E., Wakefield A.J.C., J. Chem. Soc. Chem. Comm. 969 (1984).
8. Pelzer G., DePauw E., Dung D.V., Marien J., J. Phys. Chem. $\underline{88}$, 5065 (1984).
9. Javanaud C., Eagles J., Org. Mass Spectrom. $\underline{18}$, 93 (1983).
10. Cerny R.L., Sullivan B.P., Bursey M.M., Meyers T.J., Anal. Chem. $\underline{55}$, 1954 (1983).
11. Chait B.T., and Field F.H., Proceedings of the 33rd Annual Conference on Mass Spectrometry and Allied Topics, San Diego, CA, pp. 577 (1985).
12. Benninghoven A., Surf. Sci. $\underline{35}$, 427 (1973).
13. Cooks R.G., Busch K.L., Int. J. Mass Spectrom. Ion Phys. $\underline{53}$, 111 (1983).
14. Unger S.E., Day R.J., Cooks R.G., Int. Mass Spectrom. Ion Phys. $\underline{39}$, 231 (1981).
15. Hsu B.H., Xie Y.-X., Busch K.L., Cooks R.G., Int. J. Mass Spectrom. Ion Phys. $\underline{51}$, 225 (1983).

Ion Bombardment MS: A Sensitive Probe of Chemical Reactions Occurring at the Surface of Organic Solids

B.T. Chait

The Rockefeller University, 1230 York Avenue, New York, NY 10021, USA

Organic chemical reactions occurring in the solid state or at a gas-solid interface are of interest in a number of fields including solid state polymerization and other chemical syntheses, the long-term stability of drugs, the degradation of plastics, catalysis, and a range of analytical chemical procedures. We have recently investigated the use of energetic heavy ion bombardment mass spectrometry as a new technique for monitoring microscale chemical reactions in films of nonvolatile organic solids with thicknesses ranging from submonolayers to $\sim 10^3$ molecular layers. The investigations were performed using a ^{252}Cf fission fragment induced ionization time-of-flight mass spectrometer [1] (bombardment with $\sim$100 Mev fission fragments) and a pulsed ion bombardment time-of-flight mass spectrometer [2] (bombardment with $\sim$30 keV ions), both constructed at the Rockefeller University.

The technique involves the following steps: (1) A thin film of the organic compound of interest is deposited on a metallic substrate. (2) The film is then analyzed (essentially nondestructively) by ion bombardment mass spectrometry. (3) The same film is then exposed to the reagent(s) of interest, which causes reaction to occur. (4) The chemically modified film is then reanalyzed mass spectrometrically. Differences in the mass spectra prior to and after reaction of the film gives detailed information regarding the extent of reaction and the identity of the reaction products.

To establish the generality and to investigate some applications of the procedure, the surface reactions of several organic solids were investigated with three major reagent classes:

1. <u>LIGHT REACTIVE GASES</u>

Unsaturated fatty acid salts when exposed to ozone undergo cleavage at the double bond(s) to yield aldehyde and carboxylic acid products [3]. Thus, for example, Fig. 1 shows the high mass portion of the mass spectrum of the sodium salt of vaccenic acid [$CH_3(CH_2)_5CH = CH(CH_2)_9COONa$] prior to and after exposure of the thin film to ozone. Prior to exposure, the spectrum is dominated by the $(M+Na)^+$ ion at m/z 327 as shown in Fig. 1(a). After exposure of the sample to a 1% O_3 in O_2 mixture for a time of 2 sec. the spectrum shown in Fig. 1(b) is obtained. The $(M+Na)^+$ peak has disappeared, indicating complete reaction of the starting material, and is replaced by three new peaks corresponding to reaction products of the ozone with the fatty acid salt. The product peaks which occur at m/z 245 ($OHC(CH_2)_9COONa.Na^+$), m/z 261 ($HOOC(CH_2)_9COONa.Na^+$), and m/z 283 (the $(M+2Na-H)^+$ ion corresponding to the m/z 261 ion) unambiguously define the position of the double bond. The mass spectra subsequent to ozone exposure of a series of monounsaturated positional isomers of vaccenic acid as well as of a set of polyunsaturated fatty acid salts (containing as many as four double bonds) similarly yield the position(s) of all the double bonds present. The present surface reaction probe thus appears to constitute a rapid, sensitive and unambiguous technique for determining double bond position(s) in nonvolatile olefinic compounds.

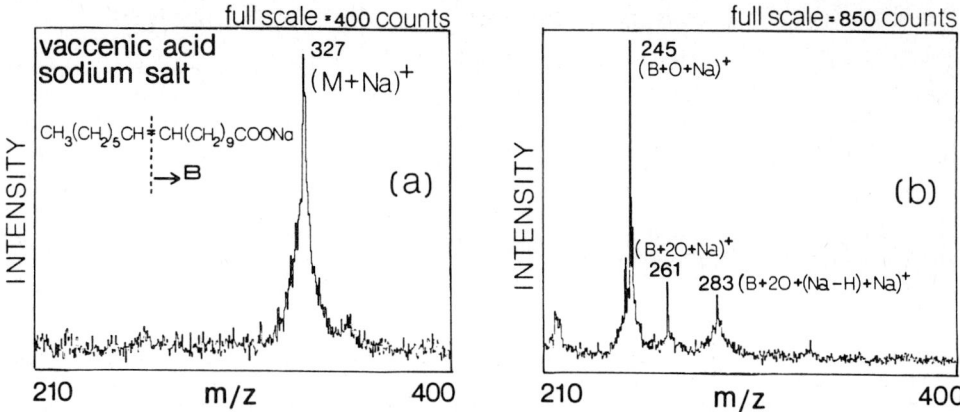

Fig. 1 High mass portion of the ^{252}Cf mass spectrum of vaccenic acid sodium salt (a) prior to and (b) after exposure of the sample film to a 1% O_3 in O_2 mixture for 2 sec.

Studies of the reaction kinetics for the ozonolysis reaction showed marked differences in the reaction rate for cis- and trans-stereoisomers of fatty acid salts. The trans-isomers had considerably lower rate constants for reaction with ozone than did the corresponding cis-isomers. We hypothesize that the different rate constants arise largely from differences in the molecular packing of the stereoisomers in the solid.

Large differences were also observed in the yields of the cluster ion species, $(nM+Na)^+$ where n=2-30, for the cis- and trans-stereoisomers of monounsaturated fatty acids salts. The trans-isomers exhibit a much higher propensity for cluster formation than do the cis-isomers. We hypothesize that the differences in the cluster yields also arise from differences in the molecular packing of the stereoisomers in the solid.

Deductions concerning the mode of formation of the cluster ions [$(nM+Na^+)$ where n=2-30] could also be made by comparing cluster yields prior to and after partial ozonolysis of the fatty acid salts. The data is consistent with direct cluster emission from the surface, as opposed to formation of the clusters above the surface by statistical recombination of independent molecules ejected by a single primary ion.

Exposure of a variety of other organic compounds to O_3, NO, and NO_2 further demonstrated that the production of new materials could be sensitively followed and their masses readily identified.

2. PHOTONS

The positive ion spectrum of an unirradiated sample film of the dye crystal violet exhibits a dominant cation peak at m/z 372 and several less intense fragment ion peaks (e.g. at m/z 356 and 340) arising from successive loss of methane. Brief irradiation of the sample with visible wavelength photons in an atmosphere of O_2 with H_2O present gives rise to intense product peaks at m/z 358, 344 and 330 arising from a series of photodemethylation reactions. Irradiation of Rhodamine B under the same conditions as those for crystal violet gave analogous de-ethylation photoproducts. Irradiation of a substituted stilbene (diethyl-stilbestrol) with 254 nm photons in air yielded a major photoproduct with a molecular weight 2μ below that of the starting material.

3. COMPLEX VAPORS

To demonstrate the feasibility of carrying out and detecting on a surface film products of a complex reaction, we carried out and monitored mass spectrometrically the coupling and cleavage steps of the Edman sequencing reaction on the terminal residue of the peptide leucine-enkephalin.

The general properties of the mass spectrometric surface reaction probe are summarized as follows: Sensitivity is high; our experience is that 10^{-9} to 10^{-15} moles of material can be measured in solid phase particle bombardment mass spectrometry. Low yields (1-2%) of selected product can be detected reliably. The method has a high surface specificity. The analysis is essentially non-destructive. The mass spectrometric analyses are rapid (typically a few minutes). The present system does not allow for the measurement of volatile compounds and volatile reaction products.

We conclude that the technique has wide potential utility for the study of surface reactions of solids with a multitude of potential practical applications.

Acknowledgment

This work was supported in part by the Division of Research Resources, National Institutes of Health.

1. B.T.Chait, W.C. Agosta, F.H. Field: Int. J. Mass Spectrom. Ion Phys. <u>39</u>, 339 (1981).
2. B.T. Chait, and F.H. Field: Proceedings of 32nd Annual Conf. Mass Spectrom. and Allied Topics, May 1984, San Antonio, p. 237.
3. P.S. Bailey: "Ozonation in Organic Chemistry, Vol. 1," Academic Press New York, 1978.

Ion-Neutral Correlations Following Metastable Decay

K.G. Standing, W. Ens, R. Beavis, G. Bolbach*, D. Main, B. Schueler, and J.B. Westmore

Physics and Chemistry Departments, University of Manitoba, Winnipeg, Canada R3T 2N2

1. Introduction

An important property of linear time-of-flight (TOF) mass spectrometers is their high efficiency for detecting the products of metastable decay [1,2]. Ions that decay in the field-free drift region contribute to the parent ion peak because the centre-of-mass velocity remains constant. However, the normal TOF spectrum provides no information on metastable decay paths or rate constants. Such data can contribute to structural elucidation and understanding of the desorption process.

2. Movable Detector Assembly

Retarding grids may be used to sort out the pattern of metastable decay [2-4]. An alternative method is the use of an ion mirror, as recently reported by Della Negra and Le Beyec [5]. We have also installed an electrostatic mirror at the end of our flight tube [6]; it is a simple 45° mirror similar to that of Danigel et al [7]. Since our mirror is only ~ 1 cm deep, it is not intended to improve the resolution, but simply to separate charged from neutral particles and to direct them into different detectors. Figure 1 shows the mirror and detector assembly; it is mounted on the end of a tube which passes through an O-ring seal, so the target to detector distance may be varied from 45 to 145 cm.

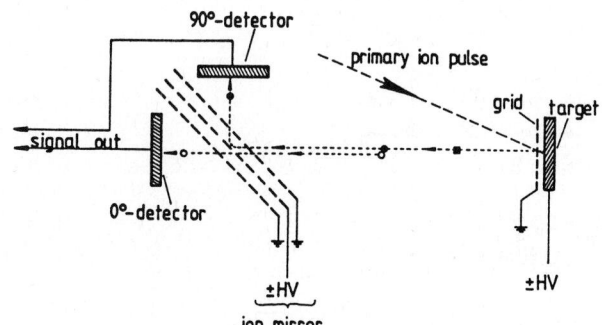

Fig. 1 Electrostatic mirror with detectors at 0° and 90°

Both detector outputs are processed simultaneously by the data acquisition system [8]. To measure correlations between charged and neutral particles, a time window is selected in the neutral spectrum. Any ion detected in the same ~ 100 μs observation period as the selected neutral (following the arrival of a given primary pulse) is recorded in a separate 2000 channel histogram. Up to four such coincidence spectra can be recorded at the same time as the normal charged and neutral spectra.

---

*Permanent address: Institut Curie, Paris, France.

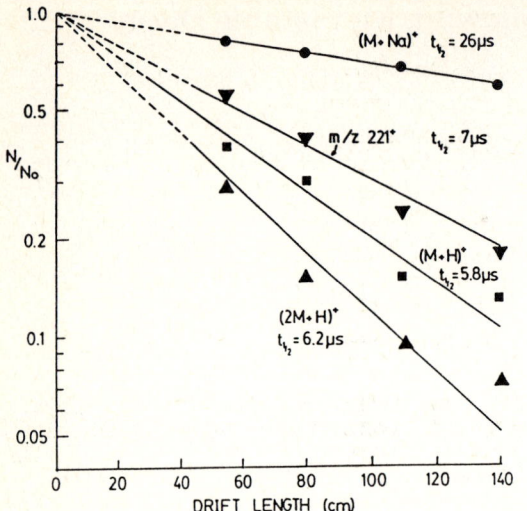

Fig. 2 Dependence of parent ion intensity on drift length for 4 peaks in the gly-gly-phe spectrum. N = number of detected parents, N_0 = number of desorbed parents.

The parent and fragment ions have almost the same velocity in the field-free region but may have quite different energies. As a result, their flight times in the field-free region will be almost the same, but the time spent in the mirror will differ. Thus the various fragment ion peaks will be separated from the parent ion peak in the charged particle detector (90°).

3. Metastable Decay of Several Peptides

Figure 3 shows TOF spectra in the molecular ion region for the tripeptide gly-gly-phe. Above is the normal positive spectrum, taken in the 0° detector with no voltage on the mirror. The peaks observed correspond to $(M+H)^+$ and

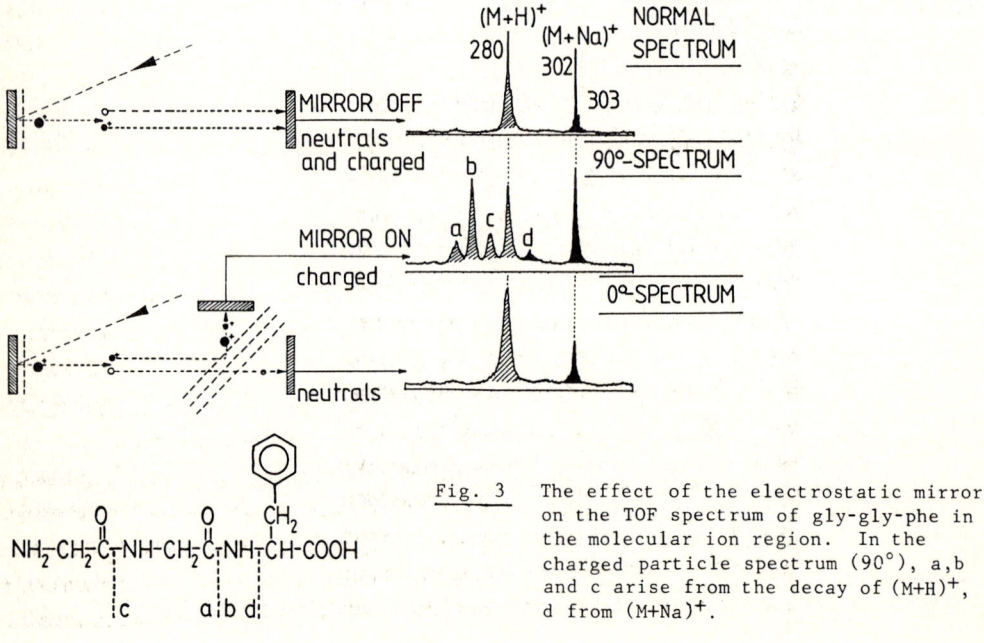

Fig. 3 The effect of the electrostatic mirror on the TOF spectrum of gly-gly-phe in the molecular ion region. In the charged particle spectrum (90°), a,b and c arise from the decay of $(M+H)^+$, d from $(M+Na)^+$.

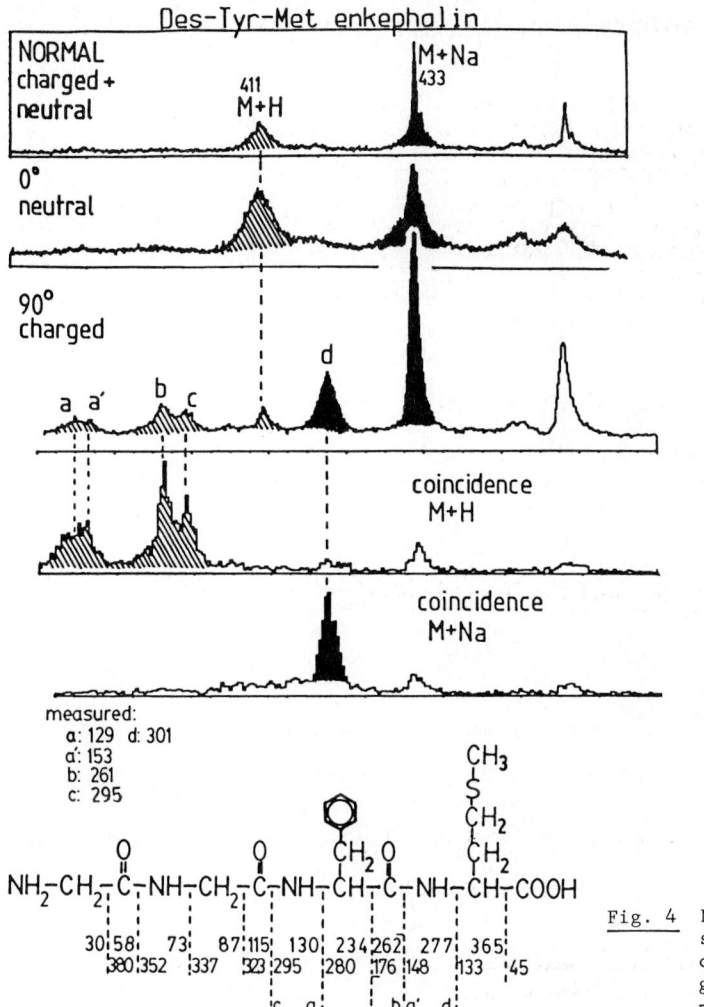

Fig. 4 Normal and coincidence spectra of the tetrapeptide des-tyr-met enkephalin = gly-gly-phe-met. The measured masses and the proposed decay paths are also shown.

$(M+Na)^+$, and the same peaks are seen in the neutral spectrum at the bottom, taken in the 0° detector with the mirror on. Additional peaks appear in the spectrum of positive ions observed in the 90° detector with the mirror on (middle spectrum); fragments a, b and c result from decay of the $(M+H)^+$ parent, while fragment d results from decay of $(M+Na)^+$.

Peaks a, b and c are consistent with fragmentation of $(M+H)^+$ at the peptide bonds, as shown. It appears that these modes of decay are completely suppressed by the presence of Na; the peak d is consistent with cleavage at the C-N bond in phenylalanine, with Na^+ remaining on the gly-gly moiety. The lifetime of the $(M+Na)^+$ ion is increased correspondingly, as shown in Fig. 2.

We have also examined the peptides gly-gly-phe-met and tyr-gly-gly-phe-met (met-enkephalin), resulting from the addition of methionine, then tyrosine, to the tripeptide above. The spectra are shown in Figs. 4 and 5. Here the

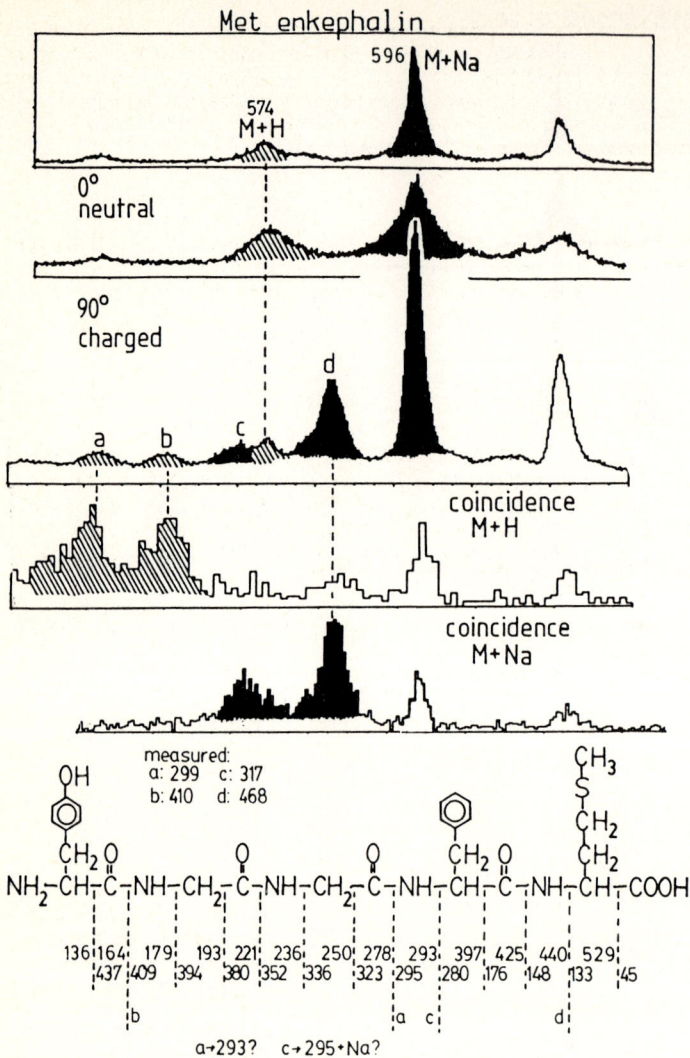

Fig. 5 Normal and coincidence spectra of the pentapeptide met-enkephalin = tyr-gly-gly-phe-met. The measured masses and the proposed decay paths are also shown.

coincidence spectra are helpful in reducing the background, and are in fact necessary to determine the parent of peak c in the met-enkephalin spectrum. Because of the limited resolution there are some ambiguities in the assignments, as indicated, but in general the spectra appear to be consistent with the same decay pattern observed for the tripeptide. The $(M+H)^+$ ion decays by cleavage at the peptide bonds, and these modes of decay are suppressed by the addition of Na. Again the decay of $(M+Na)^+$ proceeds by cleavage of a C-N bond in the C terminal amino acid, with retention of Na^+ on the N terminal fragment; for met-enkephalin an additional similar cleavage takes place within the neighboring amino acid (phe). The latter decay has also been measured by Della Negra and Le Beyec [5].

It has often been observed that $(M+Na)^+$ ions are more stable than $(M+H)^+$ ions, so it is not surprising to find that here. However, our results indicate also the mechanism for the effect; the addition of Na apparently stabilizes the N-terminal end of the molecule so that cleavage at the peptide bonds can no longer take place in the metastable decay. The molecule is then forced to decay by a path which is not visible in the $(M+H)^+$ spectrum.

4. Acknowledgments

This work was supported by grants from the U.S. National Institutes of Health (Institute of General Medical Sciences) and from the Natural Sciences and Engineering Research Council of Canada. G. Bolbach received support from CNRS (France).

References

1. B.T. Chait and K.G. Standing, Int. J. Mass Spectrom. Ion Physics 40, 185 (1981).
2. B.T. Chait and F.H. Field, Int. J. Mass Spectrom. Ion Phys. 41, 17 (1981).
3. W.W. Hunt, Jr., R.E. Huffman, and K.E. McGee, Rev. Sci. Instrum. 35, 82 (1964).
4. W. Ens, R. Beavis and K.G. Standing, Phys. Rev. Lett. 50, 27 (1983).
5. S. Della Negra and Y. LeBeyec, Int. J. Mass Spectrom. Ion Phys. 61, 21 (1984); Anal. Chem. 57, 2035 (1985); these proceedings.
6. W. Ens et al., Am. Soc. for Mass Spectrom. 33rd Annual Meeting, May, 1985.
7. H. Danigel, H. Jungclas and L. Schmidt, Int. J. Mass Spectrom. Ion Phys. 52, 223 (1983).
8. W. Ens et al., submitted to Nucl. Instr. and Meth.

Metastable Ion Studies with a ^{252}Cf Time-of-Flight Mass Spectrometer

S. Della-Negra and Y. le Beyec

Institut de Physique Nucléaire, B.P. N° 1, F-91406 Orsay, France

Time-of-flight mass spectrometry associated with a MeV ion source is certainly a good combination to investigate high molecular masses. Since the discovery of the ^{252}Cf desorption method by R. Macfarlane and co-workers [1] several P.D.M.S. systems have been built in a number of different laboratories. At our institute a time-of-flight instrument which was designed for measuring masses of exotic radioactive nuclei [2] has been used for applications in this new field as well as for new instrumental development, another ^{252}Cf instrument being available for external users. An electrostatic mirror [3] based on the Mamyrin principle [5] has been adapted to improve the mass resolution and to allow for "in flight" metastable decay studies [6]. We discuss here briefly the method and present some experimental results.

Experimental

Figure 1 shows the experimental arrangement along with a simplified diagram of the electronic set up. Each start signal given by a fission fragment allows one to record two time-of-flight spectra which we call the neutral spectrum and the reflex spectrum. A neutral fragment hitting the detector behind the mirror indicates that a fragmentation has occurred and the mass of the parent ion can be deduced from the time-of-flight measured in the neutral spectrum. The complementary charged fragment is reflected by the mirror and detected by the annular channel plates. Time correlated measurements are made to distinguish in the reflex spectrum between metastable peaks and normal peaks.

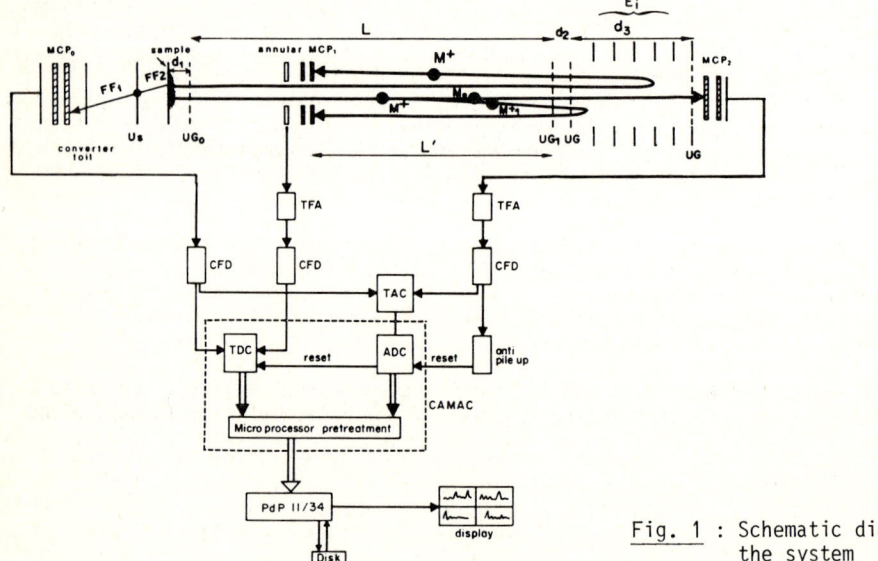

Fig. 1 : Schematic diagram of the system

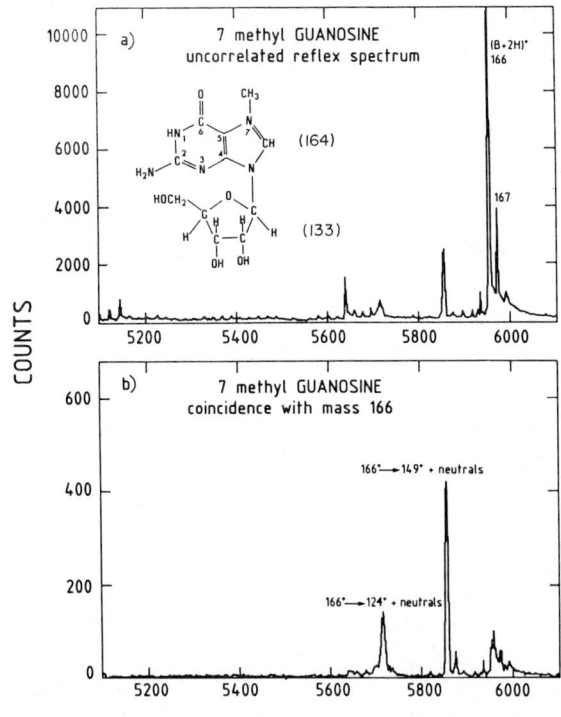

Fig. 2 :
a) uncorrelated reflex spectrum of 7 methyl-guanosine (low mass region)
b) coincidence spectrum with mass 166 on the neutral spectrum

The data acquisition system is composed of a fast microprocessor and a PdP 11/34 which had been programmed for multiparameter nuclear physics experiments. Therefore correlation between several simultaneous digital signals can be made. In particular "on line" coincidence events can be recorded with preselected windows on several parameters. In the present work we selected several time windows defined by peaks in the neutral spectrum in order to accumulate the corresponding correlated reflex spectra during data acquisition. Examples of the types of spectra obtained have already been shown for nucleosides [6]. Another simple example illustrating the method is shown in Fig. 2 with the uncorrelated total spectrum of the low mass region of 7 methyl guanosine 2a and the correlated spectrum 2b obtained when a fragmentation of the parent ion $(B+2H)^+$ = 166 occurs in flight. Here a window was set on mass 166 in the neutral spectrum. The two metastable peaks at m/z = 149 and m/z = 124 correspond to the loss of (NH_3) and ($N{\equiv}C-NH_2$) after the rupture of the bonds 1-2 and 3-4.

It must be noted that a total reflex spectrum contains all information related to "in-flight" decays whatever the ionisation methods. Coincidence techniques (either correlation or anti correlation) can be used to extract this information from the total spectrum. A similar method has also been recently applied by K. Standing and co-workers with the Manitoba time-of-flight mass spectrometer [7].

Other measurements have been made on more complex molecules [8,9]. For example Fig. 3 shows the uncorrelated reflex spectrum of the molecule met-enkephaline and the coincidence spectrum with (M+Na). The two metastable peaks at m/z 317 and m/z 464 correspond to the rupture in flight of specific bonds which have been identified [8]. The same ruptures have also been observed with the same type of molecule leu-enkephaline. New information on molecular structures can be obtained by this method since in-flight decays can be very different from prompts decays in the ionisation desorption process. Also lifetimes and internal energy release could

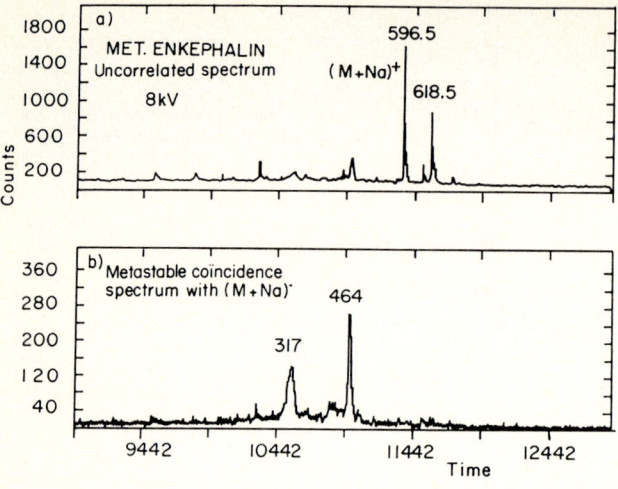

Fig. 3:
a) uncorrelated reflex spectrum of Met-enkephaline
b) coincidence spectrum with mass (M+Na) on neutral spectrum (ref.8).

be deduced in the future from this kind of measurement. The vacuum pressure is, however, an important parameter which must be carefully controlled in order to have reproducible results. For example, we have measured for a simple molecular ion $(M+H)^+$, m/z = 168 of adenosine the rate of decay in flight and/or neutralization as a function of the air pressure between 10^{-7} and 10^{-4} torr over a flight path of 80 cm. Almost 100 % neutrals were observed at $6 \cdot 10^{-5}$ torr and only a few charged fragments were observed in coincidence at this relatively high pressure. More investigations are needed, but it is certainly important for metastable ion studies with a TOF instrument to induce ion dissociations at a precise position in the field-free path region. Another important consideration is the length of the reflecting part in the mirror. In these experiments the mirror is operating at 10 KV and the total difference of time between the parent ion and the metastable ion in the reflex spectrum only depends on time differences due to their passage inside the mirror. For metastable studies only one stage reflectron is used, but to achieve a better resolution the mirror can be operated as a second order focussing reflectron with the two first grids holding 2/3 of the total voltage. Then, a mass resolution of about 3000 has been obtained for organic molecules at m/z $\sim$ 1000.

Theoretical calculations on mass resolution lead to higher values (m/Δm $\sim$ 10000) and very interesting experimental results have been recently obtained with a TOF instrument [10] for stable ions. In the case of many organic molecules however, decomposition in flight may also occur on the return flight path, and this effect can contribute to a broadening of the time-of-flight peaks.

References

[1] D.F. Torgerson, R.P. Skowronski, R.D. Macfarlane: Bioch. Biophys. Res. Commun. 60, 616 (1974).
[2] S. Della Negra, C. Deprun, H. Hungclas, H. Gauvin, J.P. Husson, Y. Le Beyec: Nucl. Instr. and Meth. 156, 355 (1978).
[3] S. Della Negra, Y.M. Ginot, Y. Le Beyec, M. Spiro and P. Vigny: Nucl. Instr. and Meth. 198, 159 (1982).
[4] S. Della Negra, Y. Le Beyec: Int. J. of Mass Spectr. and Ion Processes, 21, 61 (1984).
[5] B.A. Mamyrin, V.I. Karataev, B.V. Shmikk, V.A. Zagulin: Sov. Phys. JETP 37, 45 (1973).

[6] S. Della Negra, Y. Le Beyec: IPNO-DRE-85-01 (1985) and
    Analytical Chemistry 57, 2035 (1985).
[7] K.G. Standing et al: these proceedings.
[8] S. Della Negra, Y. Le Beyec, J.C. Tabet: IPNO-DRE-85-22 (1985).
[9] S. Della Negra, D. Fraisse, Y. Le Beyec, J.C. Tabet: Contribution to the 10th
    Int. Conf. on Mass Spectrometry, Swansea (1985).
[10] E. Nuihuis: contribution to this meeting.

Increasing Secondary Ion Yields: Derivatization/SIMS

D.A. Kidwell, M.M. Ross, and R.J. Colton
Chemistry Division, Naval Research Laboratory, Washington, DC 20375, USA

1. Introduction

When an energetic particle strikes a surface, several events happen. Primarily, the particle penetrates the surface and is implanted, but in doing so it imparts energy to the surface, which is partitioned in many ways. It is generally accepted that in metals the energy, for keV particles, is partitioned through momentum transfer collisions [1]. This cascade disrupts the surface and results in emission of electrons, neutrals and ions. Molecular SIMS is concerned with the partitioning of this energy in the surface and ultimately in the production of ions from species adsorbed on the surface. This paper discusses mechanisms of ion production from adsorbed organics. The production of positively-charged ions is emphasized, but the mechanisms proposed are just as applicable to negatively-charged species. From understanding of the mechanisms, one can optimize the production of ions and thereby increase the sensitivity of a SIMS analysis.

2. Ion Production and Ion Neutralization

During the production of ions from organics adsorbed on a surface, two uses must be made of the energy imparted during the collision cascade: the organic must be ionized and the organic must be desorbed from the surface. The order of the two events is somewhat controversial, and probably depends on the species of interest. If enough energy is imparted to the molecule by the collision cascade a radical ion can be created. Most organic molecules have ionization potentials in the range of 8-10 eV. Most metal surfaces have work functions much lower than this range. If this radical ion is produced first on the surface and then desorbed, it is energetically favorable for a surface electron to neutralize the radical cation [2-4]. This efficient neutralization mechanism for ions created on a surface produces mostly neutrals, which are not observed, and lowers the sensitivity of the analysis.

A molecule may exist on the surface associated with a metal or alkali atom or as a quaternary ammonium salt. Generally, these 'preformed ions' are observed more readily than radical cations. To observe these species, enough energy must be imparted to them by the collision cascade to be separated from their counter ions and desorbed from the surface. However, unlike a radical cation, once a preformed ion is separated from its counter ion, it cannot be neutralized efficiently by an electron from the surface.

To neutralize a quaternary ammonium ion leaving the surface, an electron must go into an antibonding orbital. This is energetically unfavorable. Hence, many of the ions survive the desorption process and are observed. Similarly, for a molecule cationized by an alkali metal, the complex ion (or preformed ion) has an electron affinity similar to that of the alkali metal ion and near the work function of the surface. In the case of cationization by a surface metal ion, these metal ions contain pi or lone-pair electrons. The electronic structure of the metal-organic complex species is significantly different from the uncomplexed molecule, which results in a lower neutralization efficiency of the complex ion.

In both cases of cationization, the similarity in electron affinity of the complex ion and the work function of the surface makes it energetically unfavorable for a surface electron to neutralize the complex ion.

One mechanism exists for the neutralization of a preformed ion: the recombination of the cation with the anion. This may be observed by varying the concentration of the counter ion (Figure 1). The effect on relative ion abundances is small for quaternary ammonium salts, but not for protonated molecules.

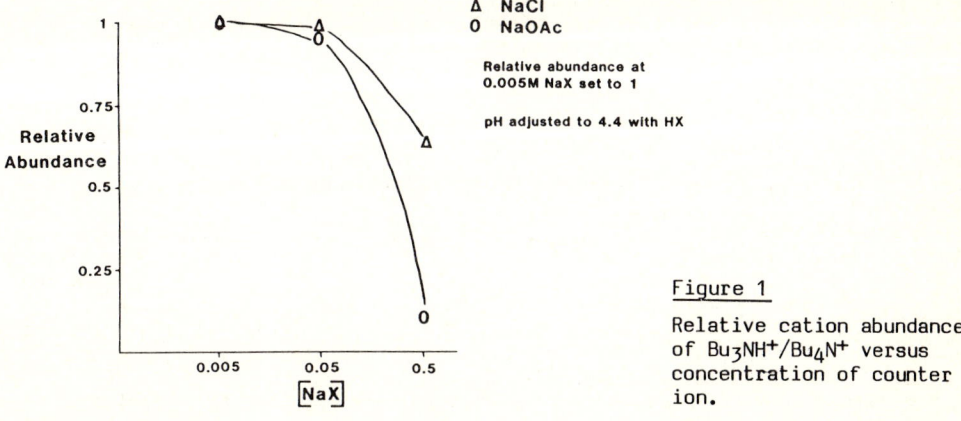

Figure 1

Relative cation abundance of Bu_3NH^+/Bu_4N^+ versus concentration of counter ion.

3. Sensitivity

The relative sensitivities of various molecular species are in the order: quaternary ammonium salts >> protonated molecules ≈ cationized molecules >> non-ionic molecules.

Besides neutralization of desorbed ions, there is one more factor favoring the higher sensitivity of preformed ions over radical cations. This consists of the amount of energy necessary to separate the preformed ion from the counter ion. As mentioned above, the ionization energy necessary to create a radical cation is about 8-10 eV. Whereas, the ionization energy of a preformed ion (energy necessary to separate the cation/anion) is less than 7 eV. Because the energy dissipates away from the impact site of the primary particle the energy available for ionization and desorption decreases with increasing distance from the impact site.

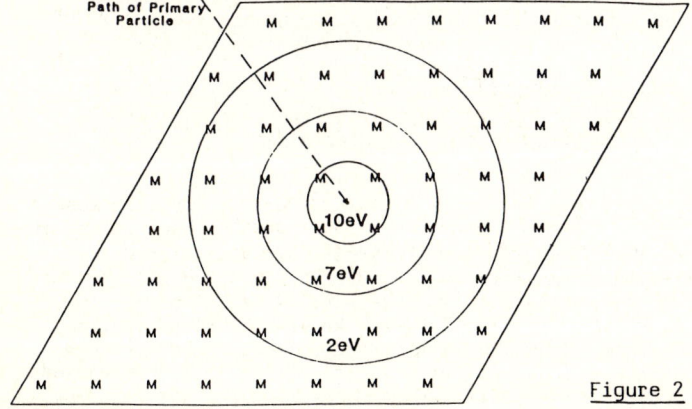

Figure 2

The available energy can be pictured as concentric rings on the surface (Figure 2). Consider three concentric rings with an energy density of 2 eV/unit area in the largest ring, 7 eV/unit area in the second ring and 10 eV/unit area in the smallest ring, nearest the impact site. Assume we have a mixture of two molecules uniformly coated on the surface, one a preformed ion with an ionization energy of 6 eV and the other a non-ionic molecule with an ionization energy of 8 eV. Also assume the energy of desorption for both molecules is 1 eV. In this highly simplified case, the preformed ion would be desorbed and ionized in the two most inner rings and be desorbed but not ionized in the outer ring. The non-ionic molecule would be desorbed and ionized only in the innermost ring and desorbed but not ionized in the two outermost rings. Comparing the number of ions produced per incident primary particle would show a higher ion yield for the preformed ion because a larger area is affected and the concentration is uniform per unit area. This is independent of any neutralization mechanism discussed above.

SIMS of an Equimolar Mixture of Four Tetraalkyl Ammonium Bromides

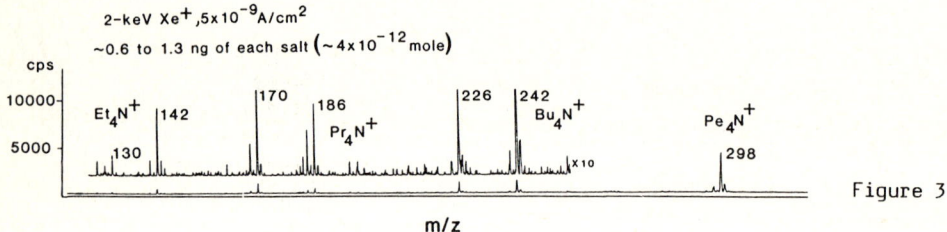

Figure 3

The energy necessary to ionize species was studied by sputtering mixtures. If one sputters an equal molar mixture of quaternary ammonium salts, one observes an increase in ionization efficiency as the size of the ammonium salt becomes larger (Figure 3). As the cation/anion distance becomes larger, the ionization energy should become smaller and a higher ionization efficiency should be observed.

4. **Sputtering Rate**

One further point needs to be examined. The higher production of ions from the larger ammonium salts may be due to a higher sputtering rate. Figure 4 shows the ion abundance decay rates for the four quaternary ammonium salts. The decay rates increased with increasing cation size. The decay rates varied by a factor of 2.5 whereas the ionization efficiencies varied by a factor of 5. The differences in decay rates may be due to several differences: ionization energies, physical sizes of the molecules, or damage effects. To separate the effects of ionization energy and size, we sputtered a mixture of tetraethyl ammonium bromide and triethylmyristyl ammonium bromide (Figure 5). The ionization energy was kept relatively constant by using similar cation-anion distances. Similar cation-anion distances were confirmed by measuring similar ionization efficiencies for both molecules. However, there is a difference in decay half-lives, which may be attributed to beam damage or differences in the molecular sizes.

5. **Effect of Primary Ion Momentum**

We have also reinvestigated the effect of momentum on ion and sputtering yields. We have separated the ionization efficiency from the sputtering yield by measuring decay half-lives. In this way, we can determine if we are observing a higher ionization efficiency because we are affecting the production of ions or increasing the number of particles desorbed, with the fraction ionized staying constant. We can also eliminate instrumental effects of changes in transmission, resolution, and extraction of ions.

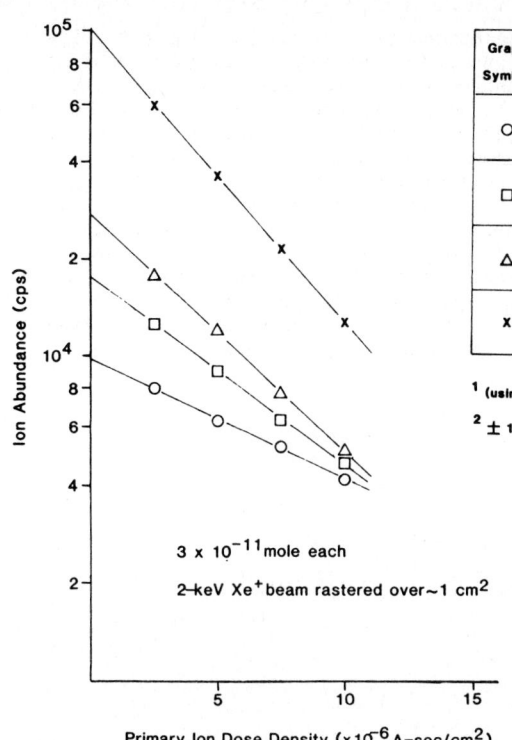
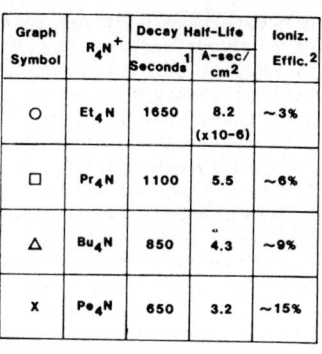

Figure 4

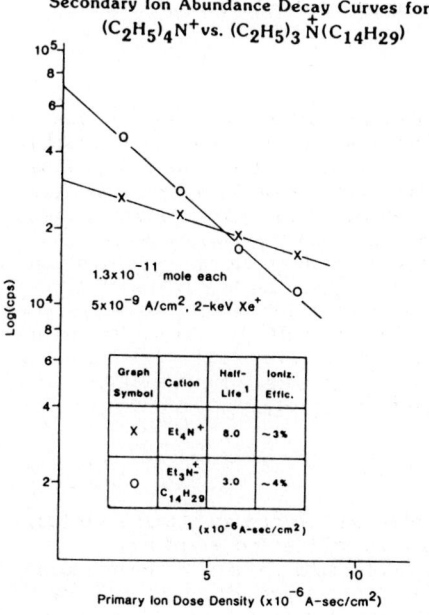

Figure 5

Figure 6

Momentum and half-life have a direct relationship (Figure 6). Figure 6 indicates that even light primary particles such as helium can give identical sputtering rates to heavier particles such as xenon. This result implies that the same area of the surface is given at least enough energy for desorption of the molecules from the surface. Where the argon and xenon data overlap, the ionization efficiencies are nearly the same because, after the primary impact event, only substrate atoms are involved in the collision cascade.

6. Summary

In summary, ionization in organic SIMS is more facile for preformed ions. The ionization efficiency increases as the size of the cation increases. The sputter rate, as measured by the decay half-life, is dependent on the momentum of the primary particle.

7. References

1. N. Winograd, Solid St. Chem., 13 285(1982).
2. N.D. Lang, Phys. Rev. B. 27 2019(1983).
3. N.D. Lang and J.K. Norskov, Phys. Scripta T6, 15(1983).
4. M.L. Yu and N.D. Lang, Phys. Rev. B 50 127(1983).
5. H.T. Jonkman and J. Michl, J. Am. Chem. Soc. 103 733(1981).

Aspects and Applications of Derivatization/SIMS

M.M. Ross, J.E. Campana, R.J. Colton, and D.A. Kidwell

Naval Research Laboratory, Chemistry Division, Washington, DC 20375, USA

1. Introduction

The sensitivity and specificity of static SIMS can be increased dramatically by using chemical derivatization to produce "preformed" ions. This sensitivity can extend to the parts-per-billion (ppb) range for selected components in complex mixtures such as body fluids. We have applied derivatization/SIMS to several classes of compounds, including peptides [1,2], saccharides [3], ketones and ketosteroids [4] and amine-containing drugs [5].

2. Why Derivatization?

Figure 1a shows a SIMS spectrum of four underivatized drugs. Protonation by an acid can be used to enhance the secondary ion yield. However, of these four drugs, methamphetamine is not observed. On the other hand, derivatizing the mixture with methyl iodide in a basic medium converts all of the amines to quaternary ammonium salts. Figure 1b shows a SIMS spectrum of this derivatized mixture. The ion abundances of the molecular ions are enhanced by at least a factor of ten over that of the protonated species. Also, methamphetamine is now observed.

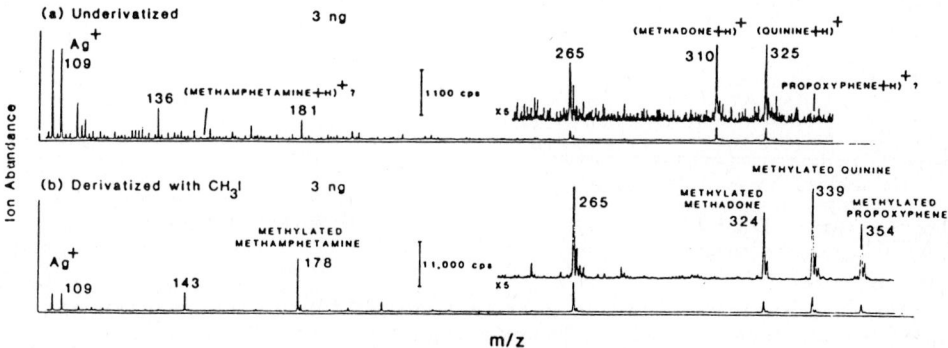

Figure 1. Human urine spiked with four drug standards.

The increase in ion abundance may be due to two factors: increased secondary ion efficiency and increased sputtering rate. Figure 2 compares the sputtering rates for the drug meperidine as its protonated and derivatized forms. The areas under the curves are proportional to the ion yields. The sputtering rates for these species are nearly identical, whereas the ion production from the derivatized meperidine is greatly increased. This indicates that the increased ionization efficiency (number of secondary ions observed/number of sample molecules deposited) of the derivatized species is responsible for the observed increase in the ion abundance.

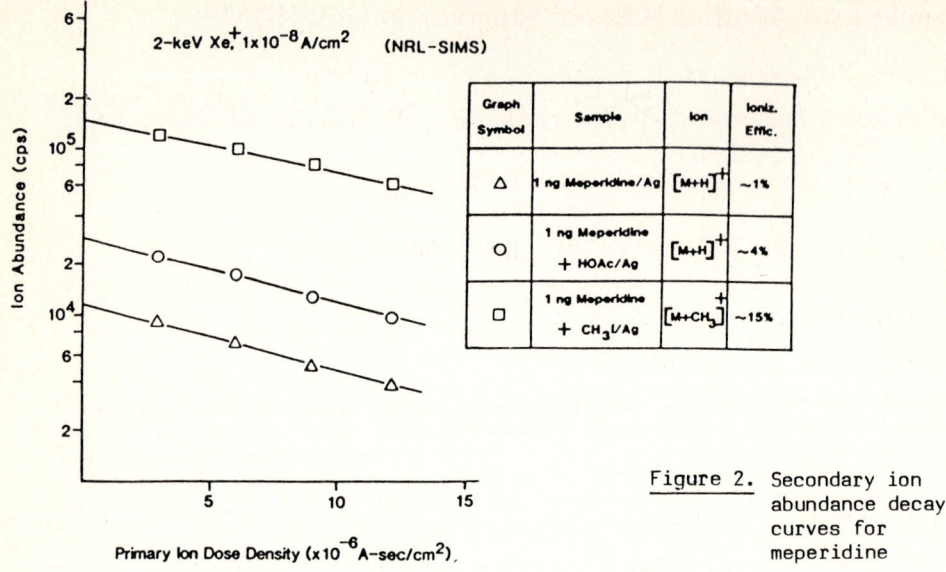

Figure 2. Secondary ion abundance decay curves for meperidine

Figure 3 shows a SIMS spectrum of an equal molar mixture of quaternary ammonium salts. These salts were produced by derivatizing benzaldehyde with four different Girard's reagents [3]. As the size of the cation increases, the ion yield also increases because less energy is necessary to separate the cation from the anion. Lower energy processes allow more efficient use of the energy of the primary particle.

Figure 3. Derivatization of benzaldehyde with an equal molar mixture of Girard's reagents

Similar size quaternary ammonium salts have similar ion yields. Mixtures of ketosteroids and aliphatic ketones derivatized with the same Girard's reagent give relative ion abundances proportional to their relative concentrations. This facilitates the quantitative analysis of mixtures of materials. Figure 4 shows the SIMS spectrum of 10 ng each of progesterone and testosterone derivatized with Girard's reagent P. Observed differences in relative ion abundances have been attributed to the different reactivities of the compounds to the Girard's reagent.

To achieve maximum sensitivity one should employ the largest quaternary ammonium salt possible. By examining simple alkyl ammonium salts, we have found that the increase in secondary ion abundance levels-off for analytes larger than tetrapentyl ammonium salt. However, other factors must be considered. For example, the chemical yield in forming larger quaternary ammonium salts decreases with longer alkyl chains.

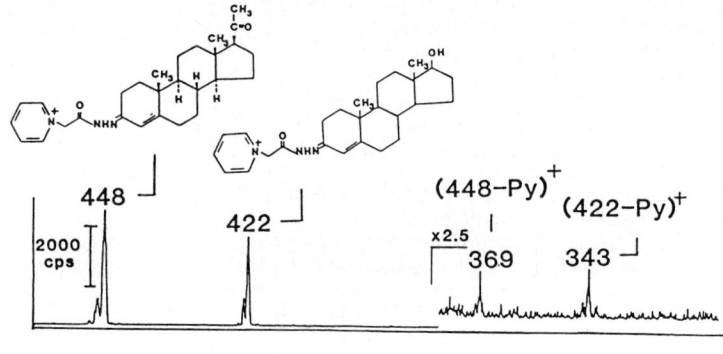

Figure 4. SIMS analysis of 10 ng each of progesterone and testosterone derivatized with Girard's reagent P

3. Sample Preparation Considerations

The detection and quantitation of other drugs such as LSD in human body fluids is a non-trivial problem due to the low dosage needed to produce a hallucinogenic response and the extensive and unknown metabolism of LSD. For the SIMS analysis of LSD a preconcentration step was necessary using a reverse-phase absorbent. Because LSD contains two potential sites for derivatization (Figure 5), most alkylating agents produce doubly-charged ions, which are seldom observed by SIMS. The reaction with methyl iodide was followed by HPLC and SIMS (Figure 6). A decrease in the ion abundance for protonated LSD was observed with time as the doubly-charged species was formed. Protontation with a weak acid is the preferred derivatization technique for LSD, particularly after the clean-up realized in the extraction.

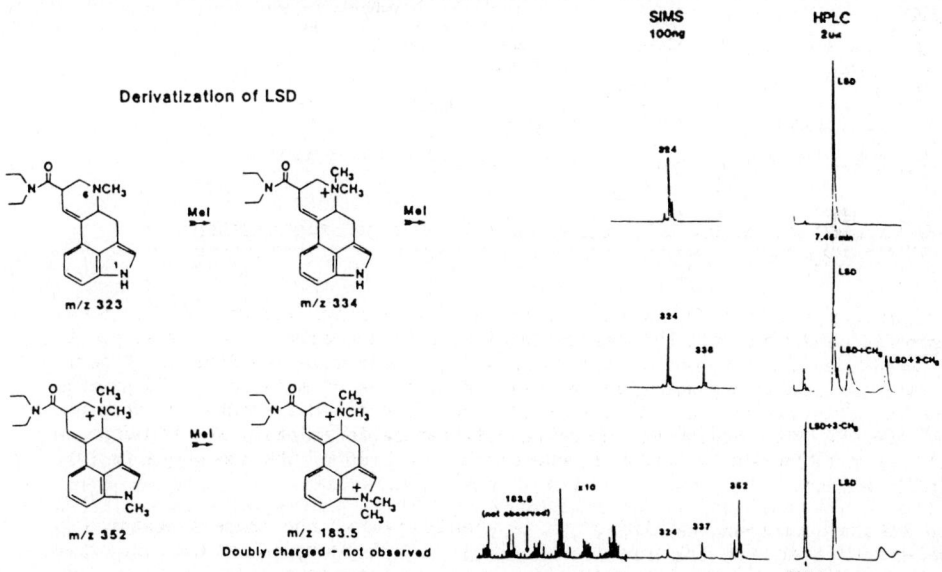

Figure 5. Derivatization of LSD

Figure 6. SIMS & HPLC analysis of methylated LSD derivatives

4. SIMS vs. FABMS

Figure 7 is a comparison of SIMS and FAB mass spectra of a mixture of five basic drugs. All of the spectra were obtained with a VG ZAB-2F mass spectrometer. Figure 7a is the FAB mass spectrum of 100 ng each of the drugs in glycerol employing the usual FAB conditions. Protonated meperidine and cocaine are observed, while the other three drugs are not observed above the background. Figure 6b is the SIMS spectrum of 10 ng each of the drugs from silver using a low current density primary beam. All of the protonated drug species except D-amphetamine are observed with good signal-to-noise ratios. Derivatization of the mixture with methyl iodide enhances the sensitivity of the analysis and allows detection of D-amphetamine, as shown in the spectrum in Figure 7c.

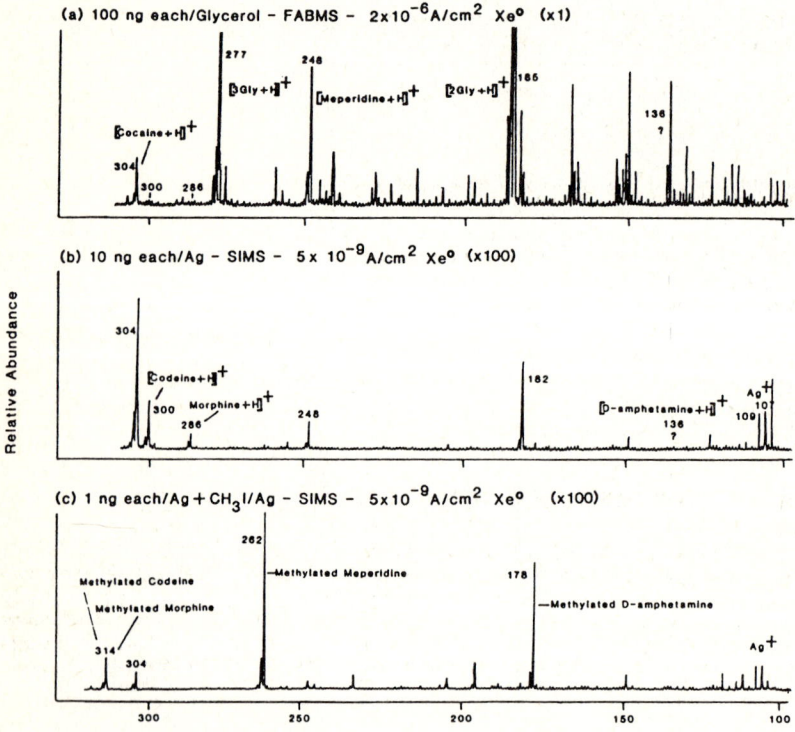

Figure 7. Comparison of FABMS and SIMS of five basic drugs

The spectra described above point out several differences between FABMS and SIMS. The FABMS method produces secondary ion currents that are approximately 100-fold greater than those from static SIMS. In addition, microgram quantities of material can be accommodated by FABMS. Another difference is the level of the background secondary ion current, which is much greater in the FAB mass spectrum due to the liquid matrix and the high primary beam current density. Therefore, although FABMS gives higher secondary ion abundances, SIMS allows analyses of smaller quantities of these drugs with superior signal-to-noise ratios. In other words, the sensitivity of SIMS is greater than that of FABMS for the analysis of these drugs.

5. Conclusions

In summary, quaternary ammonium salts make excellent SIMS labels, giving higher sensitivity compared to protonated and non-ionic compounds. As the size of a quaternary salt increases, so does the sensitivity of SIMS analysis of the salt. Indiscriminate use of derivatization can greatly decrease sensitivity, particularly if doubly-charged derivatives are formed. Use of SIMS can offer some advantages, compared to FABMS, especially for the analysis of low levels of mixture components.

6. References

1. D.A. Kidwell, M.M. Ross, R.J. Colton; J. Amer. Chem. Soc., 106, 2219 (1984).
2. D.A. Kidwell, M.M. Ross, and R.J. Colton, in Springer Ser. Chem. Phys. 36 412 (1984).
3. D.A. Kidwell, M.M. Ross and R.J. Colton, 33rd Annual Conference on Mass Spectrometry and Allied Topics, San Diego, CA, May 1984.
4. M.M. Ross, D.A. Kidwell, R.J. Colton, Int. J. Mass Spectrom. Ion Proc., 63 141 (1985).
5. D.A. Kidwell, M.M. Ross and R.J. Colton; Biomed. Mass Spectrom., 12 254 (1985).

Influence of the Target Preparation on the SI-Emission of Organic Molecules

A. Eicke[1] and A. Benninghoven

Physikalisches Institut der Universität Münster, Domagkstr. 75, D-4400 Münster, F.R.G.

[1]Present address: Institut für Physikalische Elektronik, Universität Stuttgart, Pfaffenwaldring 47, D-7000 Stuttgart 80, F.R.G.

1. Introduction

The SI-emission of organic molecules, which are deposited out of aqueous solutions on metal substrates, is strongly influenced by the chemical properties of the substrate surface, the solvent, the additives and the chemical behaviour of the substance itself /1-5/. Most often silver has been successfully applied as substrate material. In order to get informations about interactions between solved molecules and the substrate and about the composition and structure of the adsorbed organic layers, we did systematic investigations with differently pretreated silver surfaces and different solvents. The main results of the pyrimidine base cytosine are presented here.

2. Experimental

1 µl of the aqueous solution of the organic compound was deposited on 0.6 cm^2 of the Ag surface. Using concentrations of 10^{-6} to 10^{-2} mol/l the average coverage was 10^{12} to 10^{16} molecules/cm^2. For the neutral preparation with distilled water as solvent the Ag foils were etched in nitric acid and saturated with iodide or chloride in appropriate solutions. Alternatively the surface was sputtered (10^{17} Ar^+/cm^2) so that after air contact predominantly AgOH and Ag_2CO_3 was at the top layer. Using acidified solutions of cytosine the surface was iodized or chlorinated by the anions of the acids (0.1 M HCl or 0.005 M HI).

For SIMS investigations the samples were bombarded by 3 keV Ar^+ ions. To observe the temperature dependence of the SI-emission or thermally desorbed molecules (TDMS) the samples were heated at a rate of 0.7 K/s. In both methods the samples were mass analyzed in a quadrupole filter.

3. Results and Discussion

3.1 Cytosine / H_2O on AgI, AgCl and AgOH/Ag_2CO_3

The initial intensities of the characteristic ions of cytosine for different preparations are shown in Fig.1. For cytosine/H_2O on AgI the $(M+H)^+$ emission is significantly higher than the $(M-H)^-$ emission. This can be explained by the basic character of cytosine. In contrast the $(M-H)^-$ and CN^-, CNO^- emission is much higher from AgCl and AgOH/Ag_2CO_3 surfaces, which must be the result of reactions with the surface. From the negative spectra in Fig.2 we see that for these preparations without iodide a lot of cluster ions are emitted. Most of them consist of the cation Ag^- and two anions of cytosine as e.g. CN^-, CNO^-, $(M-H_2CNO)^-$ and $(M-H)^-$ and - in case of AgCl - Cl^-. Hydroxide and carbonate seem to be completely displaced from the surface, whereas on AgI surfaces cytosine is not able to form salt-like complexes with Ag or even to substitute the iodide. The results indicate that on AgCl and especially on AgOH/Ag_2CO_3 surfaces many of the molecules are chemisorbed after deprotonation or even fragmentation at basic sites. Such a complexation with the Ag

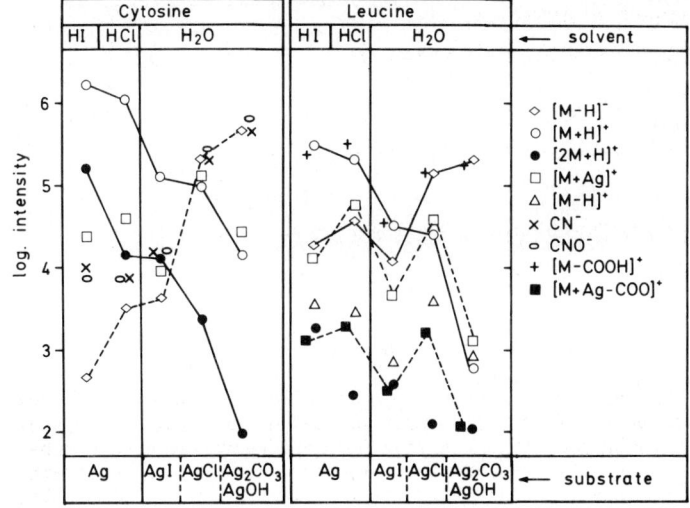

Fig.1 Intensities of characteristic SI of cytosine and leucine using different solvents and differently pretreated Ag-substrates. Coverage: 2×10^{15} molecules/cm^2. $I(Ar^+) = 0.25$ nA

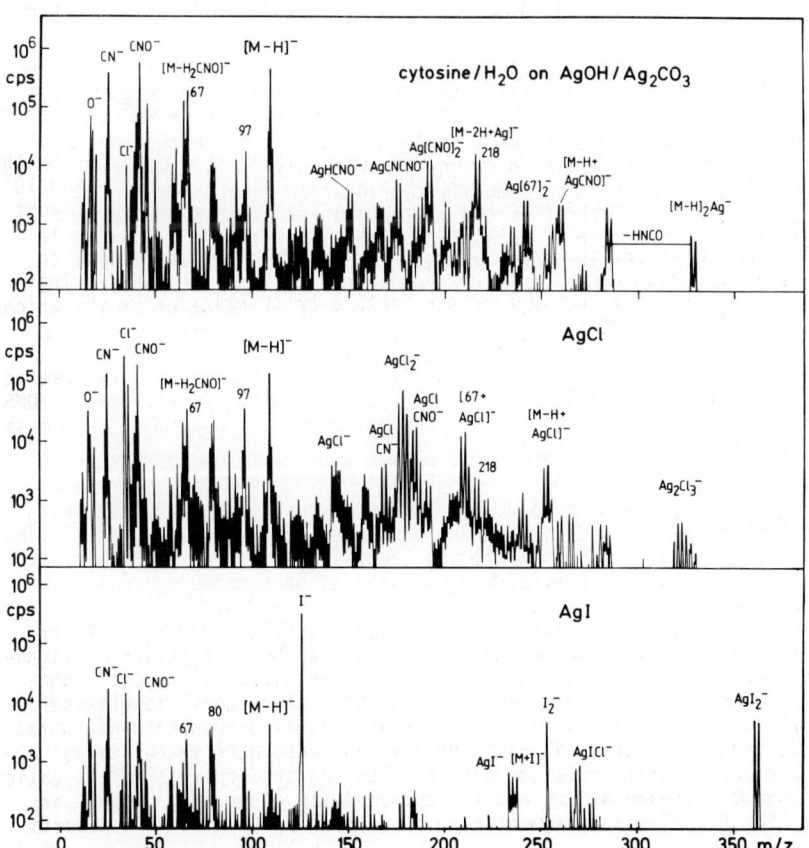

Fig.2 Negative SIMS-spectra of cytosine/H$_2$O for different Ag substrates. Coverage: 2×10^{15} molecules/cm^2

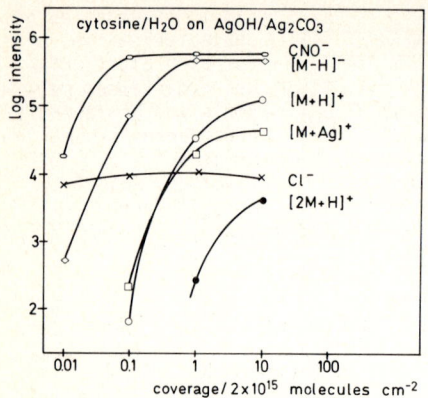
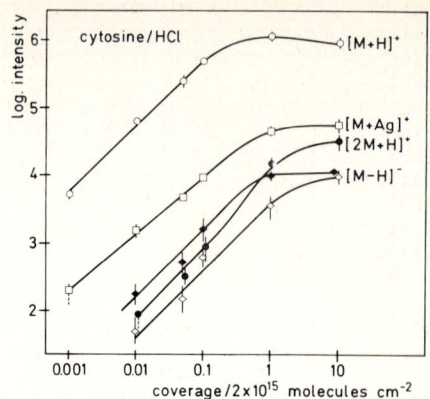

Fig.3 SI-emission of cytosine/H_2O on $\overline{AgOH/Ag_2CO_3}$ for different coverages

Fig.4 SI-emission of cytosine/HCl on Ag for different coverages

surface is also confirmed by the surface enrichment of cytosine molecules in very diluted solutions, showing a SI-emission similar to a coverage of 4×10^{14} molecules/cm^2.

The coverage dependence of the SI-emission from $AgOH/Ag_2CO_3$ which is similar to that from AgCl is shown in Fig.3. At small concentrations mainly the fragment ions CN^- and CNO^- are emitted. To higher concentrations the emission of $(M-H)^-$ and then of $(M+H)^+$ and $(M+Ag)^+$ increases drastically. This indicates that below a coverage of one monolayer equivalent the composition of the adsorbed layers is different for different coverages. This is also confirmed by the coverage dependence of the transformation probabilities, calculated from the integrated intensities of characteristic ions during ion bombardment,as shown in Table 1. Thus, at first the most reactive sites are occupied, leading to a strong fragmentation. Then the molecules are adsorbed at less reactive sites, leading to a higher emission of intact molecules. This structure becomes also evident during temperature increase: the weakly bound $(M+H)^+$ and $(M+Ag)^+$ emitting structures disappear at first, the CN^- and CNO^- intensities decrease at relatively high temperatures.

3.2 Cytosine/HCl and Cytosine/HI on Ag

Corresponding to the protonation of about 99 % of the cytosine molecules in the acidified solutions,the $(M+H)^+$ emission is dominant and very intensive (Fig.1). The $(M-H)^-$ intensities are very low. Similar to the neutral preparation,$(M-H)^-$ is lower for the preparation with iodide, whereas the emission of protonated molecules and dimers is higher.

The initial intensities of molecular ions from cytosine/HCl are proportional to the coverage up to about 10^{15} molecules/cm^2 (Fig.4). Thus, below this coverage of about 1 to 3 monolayers the composition and thickness of the organic complexes must be the same. Therefore islands exist on the surface at lower concentrations. This is also confirmed by the equivalence of the transformation probabilities of deposited molecules into characteristic ions at coverages below the saturation (Table 1). About 20 % of deposited molecules are emitted as molecular ions, most of them as protonated ones. The fraction of fragment ions is much lower than for the neutral preparation.

Informations about the composition of the organic layer are received from the temperature dependence of the SI-emission (Fig.5). $(M+H)^+$ and most of the other positive SI strongly decrease at about 390 K. The dashed line represents the integral between T and T=∞ of post ionized intact molecules, which thermally desorbed by first order from the surface. As this value is proportional to the coverage, the simultaneous decrease of the positively charged molecular SI is the result of a desorption of corresponding complexes. This desorption of the uppermost layer causes an increase of the $(M-H)^-$ emission, which also occurs during strong ion bombardment. The emission up to relatively high temperatures - similar to the neutral preparation - indicates that part of the formerly protonated molecules is strongly bound at the surface, favoring the $(M-H)^-$ emission. This fraction also seems to be proportional to the coverage, as indicated by the maximum $(M-H)^-$ intensities during temperature increase (● in Fig.4).

Table 1 Yields and transformation probabilities of characteristic SI of cytosine

	Cytosine / H$_2$O on AgCl			Cytosine / HCl on Ag		
	SI-Yield	Transf. Probability		SI-Yield	Transf. Probability	
Coverage/cm^2	2×10^{15}	2×10^{14}	2×10^{15}	2×10^{15}	10^{14}	2×10^{15}
$(M+H)^+$	0.6	-	0.5 %	7.5	16 %	17 %
$(M-H)^-$	1.3	10 %	7.5 %	0.02	0.5 %	0.5 %
$(M+Ag)^+$	0.8	0.3 %	0.8 %	0.3	3.4 %	3.7 %
$(2M+H)^+$	0.02	-	-	0.1		0.6 %
CN$^-$	1.3	30 %	10 %			
CNO$^-$	1.7	30 %	9 %			
sum of molecular ions		10 %	8 %		20 %	22 %
sum of fragment ions		≥ 60 %	≥ 20 %		< 10 %	< 10 %

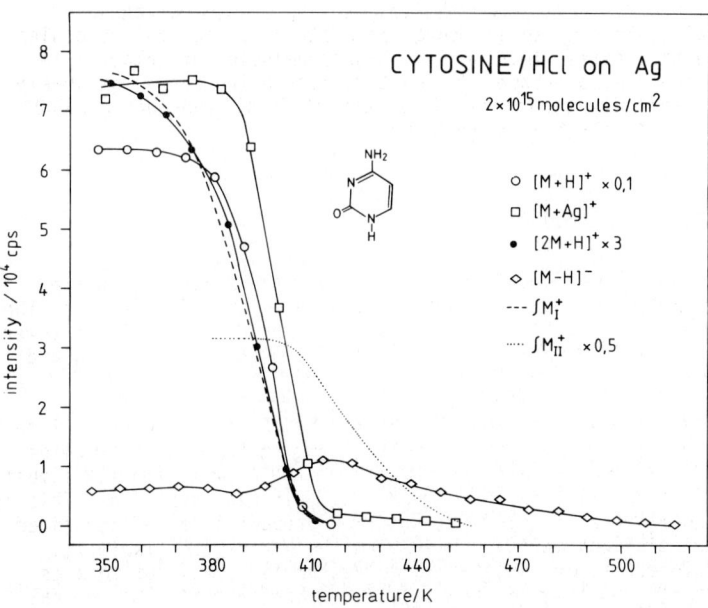

Fig.5 Intensities of molecular secondary ions and desorbed molecules during temperature increase of cytosine/HCl on Ag. $\int M^+$ corresponds to the integral of postionized gas phase molecules of cytosine between T and T=∞

The following scheme summarizes the results about the structure of the adsorbed layers of cytosine/HCl on Ag:

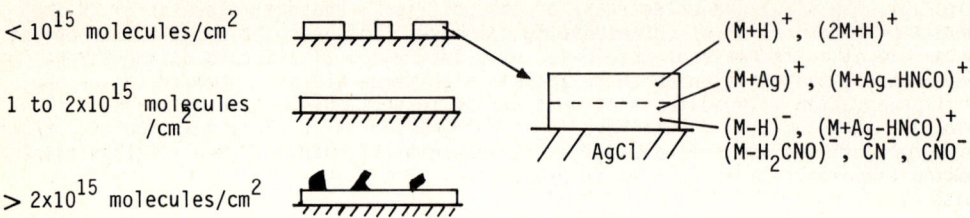

The thermally stable, chemisorbed (M-H)⁻ emitting structure does not exist after the preparation with HI. This becomes evident from the temperature dependence of the SI-emission (Fig.6): the (M-H)⁻ emission disappears simultaneously with all other molecular ions. An increase of (M-H)⁻ at higher temperatures - as is found with HCl - is not observed.

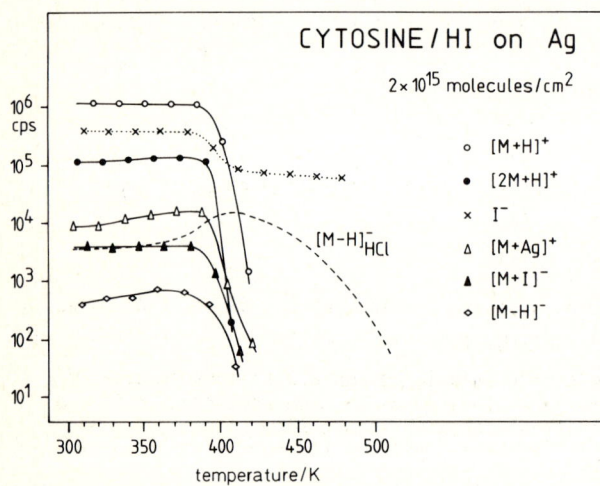

Fig.6 SIMS-intensities of molecular ions of cytosine/HI on Ag during temperature increase. $(M-H)^-_{HCl}$ corresponds to the cytosine/HCl preparation

3.3 Results of Other Compounds

Systematic investigations were also performed with the amino acid leucine. Despite the structural differences to cytosine there are strong similarities concerning the influence of the solvent and the pretreatment of the Ag-substrate on the SI-emission. For the neutral preparations of 2×10^{15} molecules/cm² (Fig.1) the (M-H)⁻ intensities again increase from AgI to AgCl to AgOH/Ag$_2$CO$_3$, the (M+H)⁺ emission decreases. The highest emission of protonated molecules is detected for preparations with acidified solutions. During temperature increase the (M+H)⁺ and (M+Ag)⁺ intensities disappear for all preparations between 350 and 390 K. This also holds true for the main part of the rather low (M-H)⁻ emission from iodic preparations. In contrast, for leucine/HCl on AgCl the (M-H)⁻ emission increases by a factor of 3 to 5 during the decrease of the positive molecular ions and disappears - similar to the very high (M-H)⁻ emission from leucine/H$_2$O on AgCl - at significantly higher temperatures (> 450 K). The detailed investigations indicate that this high (M-H)⁻ emission is the result of a deprotonation of the NH$_3^+$ group of the zwitterions and adsorption at reactive sites on the AgCl surface.

The stronger reactivity of chlorinated in comparison to iodized silver becomes also evident for adenosine monophosphate. Using HI as solvent the $(M+H)^+$ emission and also the signal to noise ratio is a factor of 3 higher compared to the acidification with HCl, whereas the emission of deprotonated and by Ag^+ cationized adenine molecules - the dominant characteristic fragment ions besides $(adenine+H)^+$ - is by this factor lower.

For Pt-oxinate, a metal chelate, the effect is more dramatically /6/. No molecular ions are emitted after deposition of some monolayer-equivalents on $AgOH/Ag_2CO_3$, but a high emission of oxine$^-$ ions is observed. From iodized surfaces the compound can be detected by significant peaks of positive molecular ions; the emission of fragment ions is much lower.

4. Conclusion

Final statements about the influence of the sample preparation on the SI-emission of polar organic compounds from Ag and some possible explanations can be summarized as follows:

i) The influence of the solvent:
The emission of certain ion species is increased, if they are preformed in the solution. This is most significant if interactions between the substrate surface and the organic molecules are very weak.

ii) The influence of the substrate pretreatment:
The reactivity of Ag-substrates, i.e. deprotonation and fragmentation of molecules increases from AgI to AgCl to $AgOH/Ag_2CO_3$. A possible explanation of this behaviour is the increase of the ionic character and by this of the acid-base strength and the solubility in water. The high $(M-H)^-$ emission can then be explained as a result of acid-base reactions (A = anion):

$$M + Ag^+ A^-(s) \longrightarrow Ag^+ (M-H)^-(s) + HA$$

Reactions of this kind were also observed between gaseous organic molecules and e.g. silver oxide /7/.

iii) The results indicate that for a high emission of molecular ions in case of preparations in the monolayer range these ions must be preformed in the solution or by interactions (reactions) with the substrate surface.

Acknowledgement

We thank W.K. Sichtermann for valuable discussions.

References

1 A. Benninghoven: Springer Series in Chem. Phys. <u>25</u>, 64 (1983)
2 K.L. Busch, S.E. Unger, A. Vincze, R.G. Cooks and K. Keough: J. Am. Chem. Soc. <u>104</u>, 1507 (1982)
3 W. Sichtermann and A. Benninghoven: Int. J. Mass Spectrom. Ion Phys. <u>40</u>, 177 (1982)
4 M.M. Ross, R.J. Colton: Anal. Chem. <u>55</u>, 150 (1983)
5 A. Eicke, V. Anders, M. Junack, W. Sichtermann, A. Benninghoven: Anal. Chem. <u>55</u>, 178 (1983)
6 B. Wenclawiak, A. Eicke, W.K. Sichtermann, A. Benninghoven: submitted to Anal. Chem.
7 M.A. Barteau, R.J. Madix: Surface Sci. <u>115</u>, 355 (1982)

Secondary Ion Formation Processes in Amino Acid - Metal Adsorption Systems

D. Holtkamp, M. Kempken, P. Klüsener, and A. Benninghoven

Physikalisches Institut der Universität Münster, Domagkstr. 75, D-4400 Münster, F.R.G.

1. Introduction

Due to its high sensitivity, especially by using a time-of-flight instrument, secondary ion mass spectrometry has been applied in several fields of biological and medicinal research /1/. However, only little understanding concerning the ion formation process has been obtained so far, mainly caused by the very complex surface composition which results from the conventional sample preparation out of a solution. The large number of possible interactions in such samples makes it difficult to get reliable information about the influence of the substrate on molecular secondary ion emission. Thus, well-defined adsorption systems are required.

We have approached this problem by using a molecular effusion technique to prepare amino acid overlayers with a controlled thickness under UHV conditions /2/. Regarding the protonated/deprotonated ion emission we found a strong influence of the substrate /3,4/: from Au and Pt both ions are found for all coverages while Ni yields only $(M-H)^-$ in the submonolayer range. Combined SIMS/TDMS experiments /5/ revealed that the proton transfer leading to the formation of $(M+H)^+$ is only possible between weakly bound amino acid molecules. In contrast to other adsorption systems which are characterized by the emission of positive metal-containing cluster ions (Ni - H : Ni_2H^+ /6/, Ni - CO : $NiCO^+$, Ni_2CO^+ /7/, Ag - CH_3OH: $Ag(CH_3OH)_n^+$ /8/), amino acid - metal systems yield both MeM^+ and MeM^- with comparable intensities /9/.

In this work, the characteristic features of spectra from evaporated amino acids on metals and a more detailed investigation of the coverage dependence of secondary ion emission from Gly on Au are reported. In order to obtain more information about the formation of $(M\pm H)^\pm$, coadsorption experiments of Gly-d_5 and Phe were carried out. The experimental arrangement has been described in ref.5. All experiments were performed at a substrate temperature of 273 K and under static ion bombardment conditions.

2. Secondary ion emission from amino acid-metal systems

The main features of secondary ion spectra from in situ prepared amino acid overlayers on metals are illustrated in fig.1, showing positive and negative spectra of Ala on Au. Due to the clean preparation method there are no contaminations, i.e. alkali ions or salts, present. The base peaks are the protonated and deprotonated molecule and the characteristic fragment $(M-COOH)^+$. The ions $(M-H)^+$ and $(M-3H)^-$ are unique for monolayers of amino acid on Au and Pt, while $(2M\pm H)^\pm$ are found from multilayers as well. Although the

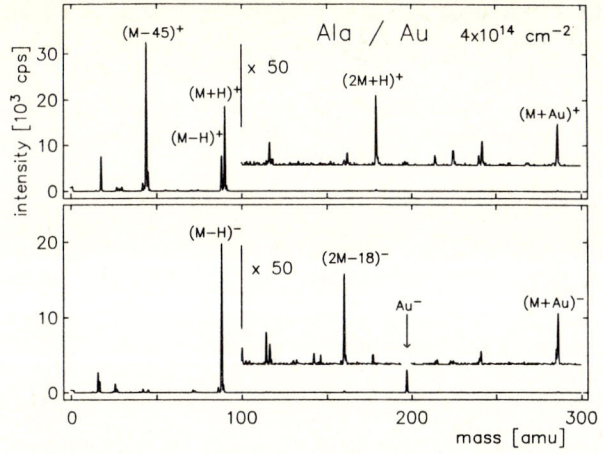

Fig.1 Positive and negative secondary ion mass spectra of $4 \cdot 10^{14}$ cm^{-2} Ala evaporated onto Au (Ar$^+$, 3.5 keV, I_p=6 nA/cm^2)

ratio of the Au$^-$ to Au$^+$ yields is more than a factor of 10^3, (M±Au)$^±$ are emitted with comparable intensities. Both cationized and anionized molecular ions are also found for other substrates /9/. The negative ion at m/z = 160 results from fragmentation of the amino acid dimer. This species cannot be ascribed to dipeptide formation in the molecular beam source as experiments with dipeptides evaporated onto Au show both protonated and deprotonated ion emission.

The structures of (M-H)$^+$ and (M-3H)$^-$ were determined by comparing the spectra of Gly-n, Gly-2,2-d$_2$ where the carbon-bound hydrogen atoms are replaced by deuterium, and the completely deuterated species Gly-d$_5$. Both Gly-n and Gly-2,2-d$_2$ yield (M±H)$^±$ signals, while Gly-d$_5$ is characterized by (M±D)$^±$. (M-1)$^+$ is shifted to (M-2)$^+$ for both deuterated compounds and (M-3)$^-$ to (M-4)$^-$ for Gly-2,2-d$_2$ and (M-6)$^-$ for Gly-d$_5$. From these results one can conclude, that both ions have lost one carbon-bound hydrogen, and the negative ion additionally 2 hydrogens from the functional groups.

The coverage dependence of the molecular secondary ion intensities has been reported earlier for the system Gly on Ni /9/ and for

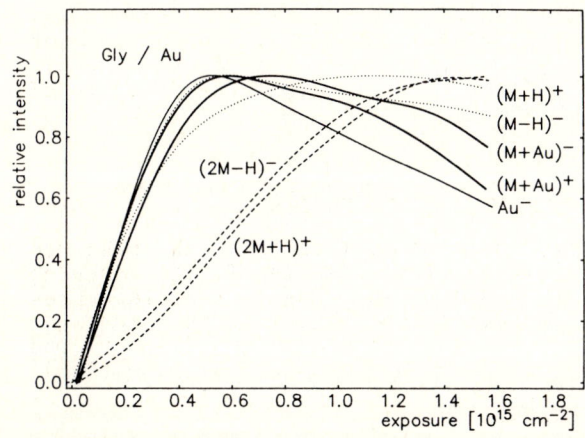

Fig.2 Change of molecular secondary ion intensities of Gly on Au during layer growth

the ions $(M\pm H)^\pm$ and Me from Gly on Au /4/. The latter system has been investigated in more detail, and results are presented in fig.2. In contrast to Gly on Ni, all molecular ions are detected for coverages below $4 \cdot 10^{14}$ cm^{-2}. Apart from $(2M\pm H)^\pm$ all ion intensities increase similarly in the submonolayer region, show a distinct maximum, and then decrease with different slopes. The $(2M\pm H)^\pm$ signals increase significantly slower and reach maximum intensity much later.

3. Hydrogen exchange reactions

Coadsorption experiments of Gly-d_5 and Phe on different substrates revealed that significant exchange reactions of hydrogen and deuterium are observed in multilayers independent of the substrate, and for monolayer coverages on Au and Pt, but not on Ni. This allows the conclusion, that substrate contact is not necessary, but it may prevent exchange.

The upper part of fig.3 compares the spectra of Gly-d_5 on Au before and after coadsorption of Phe. At masses below the characteristic ions new lines appear, indicating hydrogen exchange. The maximum mass shifts are 2 for (M-COOD)$^+$ and (M-D)$^-$ and 3 for (M-D)$^+$. The maximum mass shift of (M+D)$^+$ cannot be determined because it interferes with (M-D)$^+$. Corresponding spectra for Phe are shown in the lower part of fig.3. For (M-COOH)$^+$ and (M-H)$^-$ the maximum mass shifts are identical to Gly-d_5, for (M+H)$^+$, mass shifts up to 4 are found. All these shifts coincide with the number of H/D atoms in the

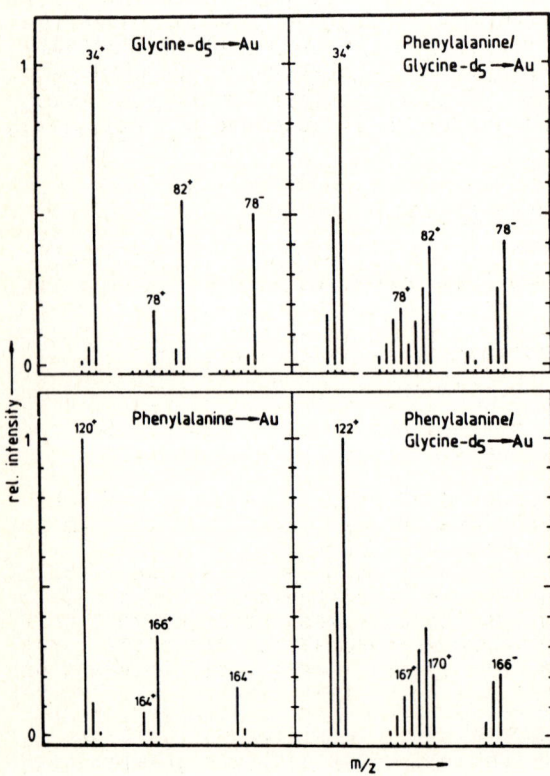

Fig.3 Normalized mass spectra from coadsorbed amino acids on Au. Upper part: Gly-d_5 before and after coadsorption of Phe. Lower part: Phe on a clean and a Gly-d_5 precovered surface. Coverage: approximately 0.5 monolayer equivalents for each compound

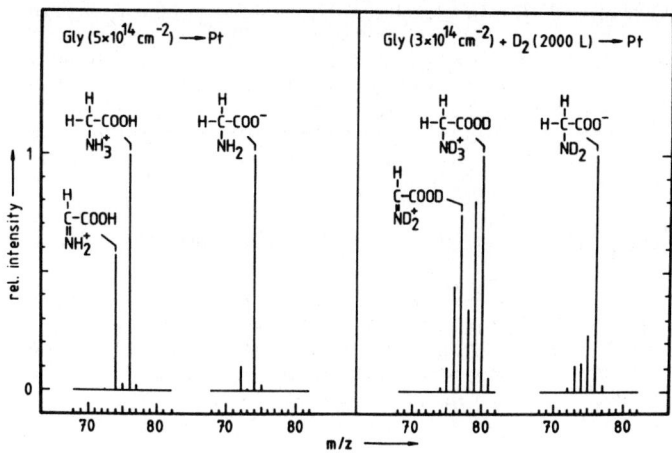

Fig.4 Relative intensity of protonated and deprotonated molecular secondary ion emission from a Gly submonolayer on clean Pt (left part) and Deuterium precovered Pt (right part) /10/

functional groups, supporting the assumption that carbon-bound hydrogen cannot be exchanged.

Coadsorption experiments of amino acids and deuterium from the gas phase pointed out that only on Pt exchange reactions take place /10/. Figure 4 shows a spectrum of Gly-n evaporated onto a Deuterium saturated Pt surface. Similarly to the coadsorbed amino acids only hydrogen from the functional groups may be exchanged. These experiments show that hydrogen exchange is only possible if the amino acid is bound weakly to the metal and deuterium is adsorbed dissociatively.

4. Model of amino acid-metal interaction

The experimental results presented previously and in this paper lead to the model of amino acid - metal interaction described in fig.5. The basis for understanding is the zwitterion structure of amino acids in condensed matter, i.e. their ability to act as proton donors and acceptors as well. The coadsorption experiments reveal that all hydrogen atoms from the functional groups are equivalent and may be involved in this process. So in multilayers free proton exchange is possible between adjacent molecules leading to the emission of the characteristic molecular ions $(M\pm H)^{\pm}$.

Both the coverage dependence and TDMS results /5/ indicate that the amino acid - metal bond on Au and Pt is similar to the amino acid - amino acid bond. Proton exchange between different species is possible and $(M\pm H)^{\pm}$ can be emitted. Additionally $(M-H)^+$ and $(M-3H)^-$ are observed, resulting from the loss of a carbon-bound hydrogen. On the other hand, the amino acid - Ni bond is much stronger /9/ prohibiting the formation of $(M+H)^+$ from submonolayers. After the first layer is filled completely, the amino acid - amino acid interaction becomes more important and $(M+H)^+$ may be emitted.

The formation of metal-containing secondary ions is somewhat more complicated and cannot be explained by a simple recombination

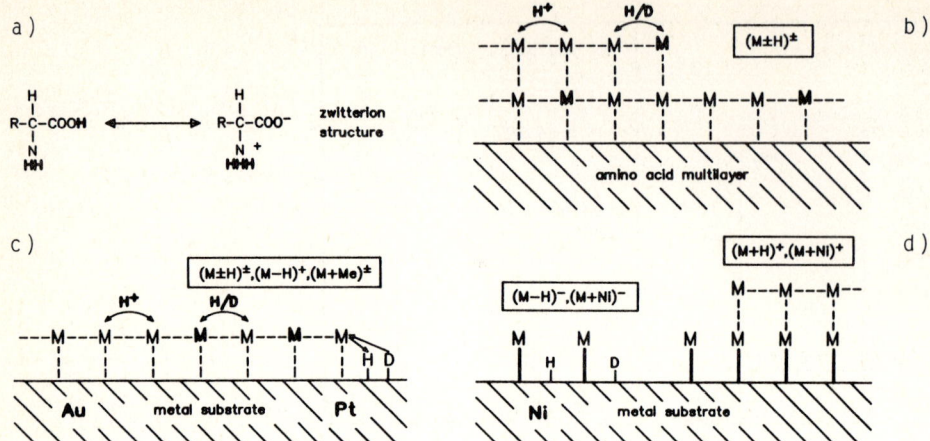

Fig.5 Model of the amino acid - metal interaction (M denotes amino acid molecule, boldfaced **M** a different species): a) Zwitterion structure of amino acids, b) Multilayer, c) Monolayer on Au and Pt, d) Monolayer on Ni

model. While both $(M+Me)^+$ and $(M+Me)^-$ are emitted from submonolayers of Gly on Au and Pt with comparable intensities and no Au^- is observed (fig. 1), only the negative species is found on Ni. The positive ion is not observed, unless the first monolayer is filled, showing the same coverage dependence as the protonated molecule /9/. As the intensity of these ions is not correlated with $Me^{\pm}$ we believe that they are formed by sputtering and subsequent decay of $Me_m M_n$ clusters. Concerning Ni, $(M+Ni)^-$ may be formed if n=1, while $(M+Ni)^+$ deserves at least n=2. The value of m certainly depends on the adsorption geometry but is unknown so far.

In conclusion we state, that amino acid overlayers prepared by evaporation onto clean metals are suitable for investigating the emission process of secondary ions from organic overlayers.

References

1 A.Benninghoven, E.Niehuis, T.Friese, D.Greifendorf and P.Steffens: Org. Mass Spectrom. 19, 346, (1984)
2 D.Holtkamp, M.Jirikowsky, W.Lange and A.Benninghoven: Appl. Surface Sci. 17, 296, (1984)
3 W.Lange, M.Jirikowsky and A.Benninghoven: Surface Sci. 136, 419, (1984)
4 A.Benninghoven, W.Lange, M.Jirikowsky and D.Holtkamp: Surface Sci. 123, L721, (1982)
5 D.Holtkamp, M.Jirikowsky, M.Kempken and A.Benninghoven: J. Vac. Sci. Technol. A 3, 1394, (1985)
6 A.Benninghoven, P.Beckmann, D.Greifendorf, K.H.Müller and M.Schemmer: Surface Sci. 107, 148, (1981)
7 A.Brown and J.C.Vickermann: Surface Sci. 117, 154, (1982)
8 R.Jede: Thesis, Münster (1985)
9 D.Holtkamp, M.Kempken, P.Klüsener and A.Benninghoven, in: SIMS V, Springer Series in Chemical Physics, (in press)
10 M.Jirikowsky, Thesis, Münster (1984)

Analytical Applications of High-Performance TOF-SIMS

W. Lange, D. Greifendorf, D. van Leyen, E. Niehuis, and A. Benninghoven

Physikalisches Institut der Universität Münster, Domagkstr. 75,
D-4400 Münster, F. R. G.

1. Introduction

Trace analysis of organic compounds offers a new field of application for mass spectrometry. Secondary ion mass spectrometry combined with a time-of-flight analyser has turned out to be an extremely sensitive technique for molecular weight determination and structure elucidation of large, involatile and thermally labile compounds.
The main advantages of a TOF-instrument are an unlimited mass range, quasisimultaneous detection of all masses and high transmission providing an efficient analysis of generated secondary ions.

A high-performance TOF-instrument, built in Münster 4 years ago, was first described in detail by P.Steffens in 1982 [1,2]. The instrument is composed of a pulsed, mass separated primary ion source, a multiple focusing analyser of Poschenrieder type and a detection system consisting of channelplate, scintillator, light guide and photomultiplier. The primary beam system produces ion pulses of less than 10nsec and an averaged current of 1 - 100 pA. The detector is operated in the single ion counting mode, SI can be postaccelerated up to 20 keV for higher detection efficiency. The typical time needed for the accumulation of a spectrum is in the range of 10 - 100 sec.

Since 1982 this TOF-SIMS has been applied for routine analysis of a wide range of organic samples such as aminoacids, PTH-aminoacids, pharmaceuticals, peptides, saccharides, oligonucleotides and synthetic polymers [3].

This paper will summarize preliminary results concerning molecular weight determination, sequence analysis, sites of ionization and detection limits of peptides.

2. Sample Preparation

Typically 0.1 μl of a 10^{-3} - 10^{-4} mol/l solution of sample molecules are deposited on an area of 1 - 10 mm^2 of an etched Ag or Au foil. This results in a stable monolayer or submonolayer of molecules in the bombarded area of 1 mm^2. Thus the optimum sample size for TOF-SIMS is in the pmol range. The minimum sample size can be estimated from transformation probabilities $P=N_d/N \cdot T$ with N_d = detectable molecular ions, N = number of molecules in the analysed area and T = transmission of the instrument. For several peptides a detection limit in the fm-range can be expected.

An alternative preparation technique is the deposition of molecules in a regenerating liquid matrix of low vapor pressure. For optimum results 1 - 10 nmol of sample material are dissolved in 1 - 2 μl of glycerol. The droplet is then placed on the same target holder used for solid SIMS. Liquid SIMS measurements can be easily performed in our TOF-instrument without further modifications.

3. Results

Figure 1 presents the positive SI mass spectrum of the tryptic peptide T9 of apolipoprotein AI [4]. The AI-protein, extracted from human blood serum, consists of 243 aminoacids with total molecular weight of 28331 amu. Tryptic cleavage results in 38 fragment peptides. SIMS spectra of these peptides, taken after HPLC separation of the reaction mixture, show high $(M+H)^+$, $(M+Na)^+$, and $(M+Ag)^+$ intensities allowing a fast identification of enzymatic degradation products.

The spectra of the nonapeptide Bradykinin (Fig. 2) were obtained from a glycerol matrix and a Ag-surface respectively under the same bombardment conditions. Both measurements provide equal molecular weight and sequence informations. Signal intensities of secondary ions differ by a factor of about 2; for a more detailed comparison of SI emission from Ag and glycerol in the mass range up to 1000 amu see Junack et al. [5].

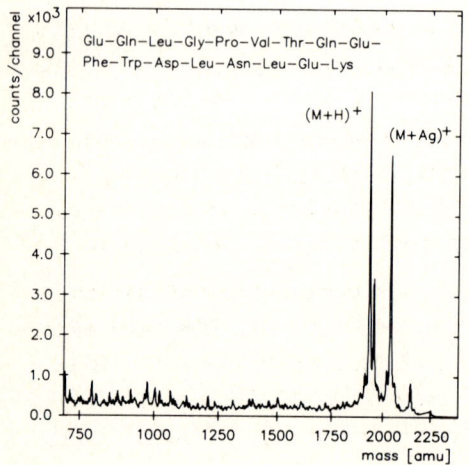

Fig. 1 Positive SIMS spectrum of tryptic peptide T9 of apolipoprotein AI after HPLC-separation; sample: 1 pmol on Ag; primary ions: $1 \cdot 10^9$, Ar^+, 12 keV

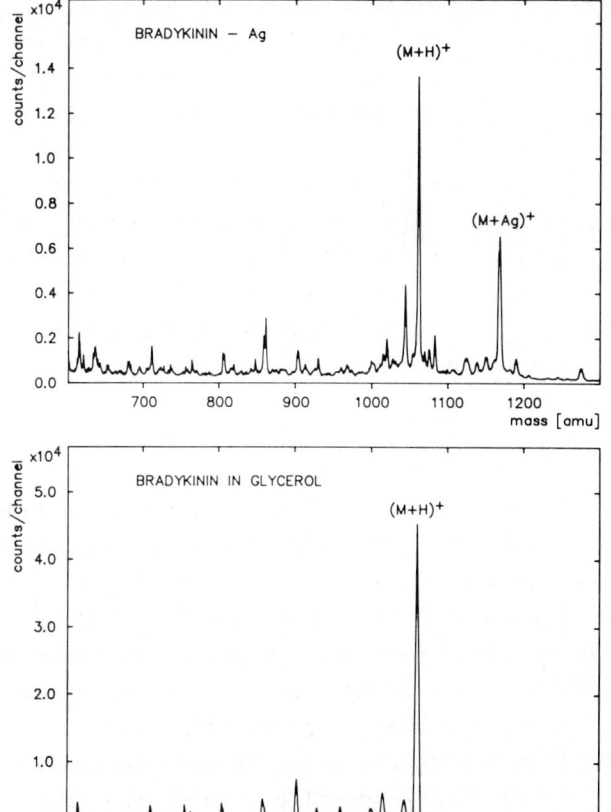

Fig. 2 Positive SIMS spectra of Bradykinin on Ag and in glycerol; sample: 1 pmol (Ag), 1 nmol (glycerol; primary ions: $1 \cdot 10^9$, Ar^+, 12 keV

3.1 Sequence Analysis

In general, sequence ion intensities of peptides in solid or liquid SIMS spectra are too low for structure determination of unknown peptides. For sequence analysis of unknown samples we combined TOF-SIMS with chemical methods such as Edman degradation or enzymatic cleavage by exopeptidases. A crucial step in peptide analysis by Edman degradation is the identification of PTH-aminoacids formed during the chemical reaction. TOF-SIMS spectra show intense $(M+H)^+$, $(M+Au)^+$, $(2M+Au)^+$ and $(M-H)^-$ signals for all PTH-aminoacid standards investigated. A reliable detection in the sub pmol range is possible [6].

Figure 3 presents the results of enzymatic digestion of Met-Lys-Bradykinin by carboxypeptidase Y. The enzyme cuts off the last aminoacid of a peptide starting at the carboxyl terminus. From the mass shift of the $(M+H)^+$ peak in the SIMS spectra, taken after different reaction times, one can determine the sequence of investigated peptides.

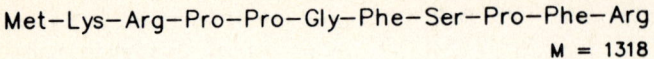

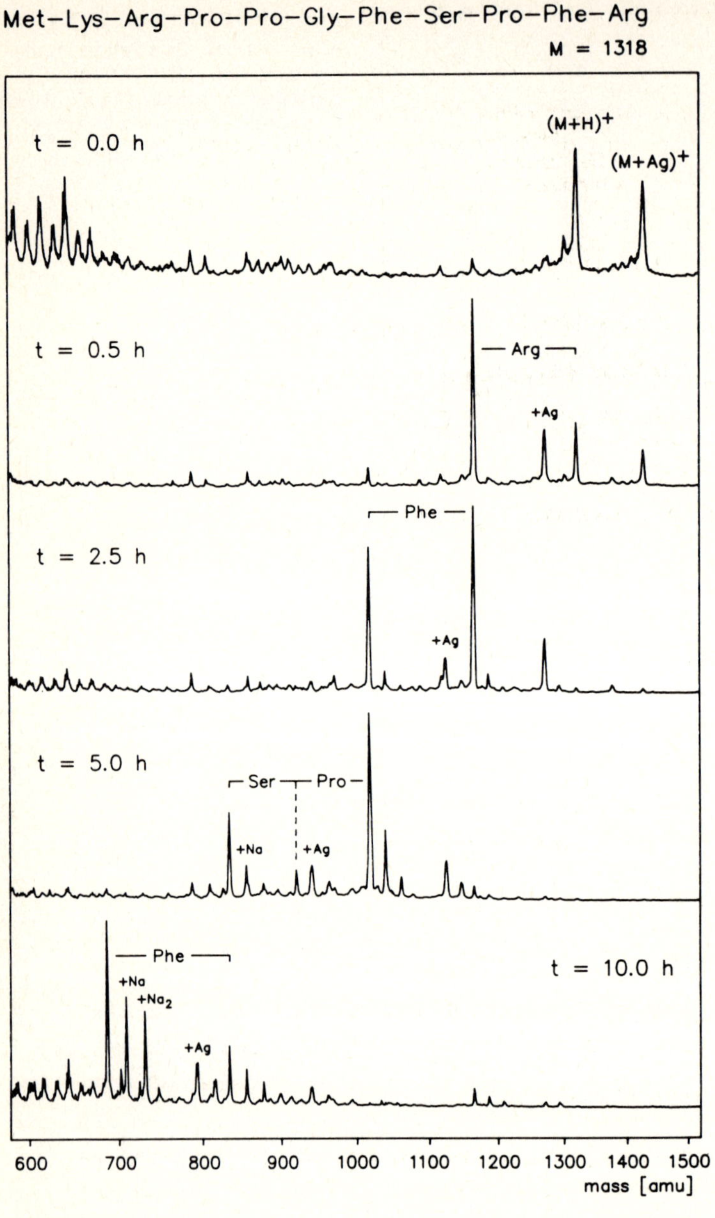

Fig. 3 Positive SIMS spectra of Met-Lys-Bradykinin + carboxypeptidase Y after different reaction times; sample: 1 pmol on Ag; primary ions: $1 \cdot 10^9$, Ar^+, 12 keV

For each spectrum an amount of 1 pmol of sample is needed. When using this technique one has to take into account that the velocity of enzymatic cleavage may vary by a factor of about 10 depending on the special aminoacid at the COOH-terminus of the peptide. Pro for instance is cut off by carboxypeptidase Y much slower than Arg or Phe, thus explaining the low intensity of the fragment peptide after release of Pro at mass number 903.

3.2 Derivatization, Influence of Reactive Groups

To get a better understanding of the sites of ionization in the peptide chain, and the influence of primary structure on secondary ion yields, we carried out several derivatization experiments and systematic comparisons of a variety of different peptides.

Molecules were derivatized by N-Methyl-bis-trifluoroacetamide (MBTFA). This reagent trifluoroacetylates $-NH_2$, $-NH$, $-SH$, $-OH$ and $-COOH$ groups, substituting a proton by a $-\underset{O}{\underset{\|}{C}}-CF_3$ group.

Spectra of partially and fully derivatized Bradykinin, a peptide containing two Arg with additional amino groups in their side chains, revealed that there is no $(M+H)^+$ formation if all NH_2 groups in the molecule are blocked by MBTFA. When the COOH terminus is also derivatized, the emission of $(M-H)^-$ ions disappears. Fully derivatized Bradykinin only shows molecular ions due to cationization by Na and Ag.

These results suggest that protonation occurs at the amino groups at the N-terminus or in the side chains of basic aminoacids Arg and Lys. Deprotonation seems to be correlated with the COOH groups, although additional sites of deprotonation cannot be excluded by our measurements. SIMS spectra of Met-Enkephalinamide, for instance, show $(M-H)^-$ ions although there is no free carboxyl group in the molecule. Finally cationization by Na or Ag seems to be independent of reactive groups in the peptide.

When comparing secondary ion emission of different peptides containing up to 5 basic aminoacids Arg and Lys it became obvious that the number of amino groups has a major influence on relative and absolute $(M+H)^+$ ion yields.

Peptides without an Arg or Lys, the enkephalins for instance, show high cationized SI signals with relative low $(M+H)^+$ intensities. Peptides with a single Arg or Lys reveal maximum yields of $(M+H)^+$, see for example Des-Arg-Bradykinin or tryptic peptides of apolipoprotein AI.

Absolute molecular ion yields decrease by a factor of about 3 - 10 with an increasing number of amino groups within the peptide. This result might be explained by a strong interaction of NH_2-groups with the etched Ag-surface. The effect can be demonstrated in the spectra of enzymatic degradation of Met-Lys-Bradykinin (Fig. 3). The molecule contains 3 basic aminoacids. After the first reaction step cutting off the C-terminal Arg, the $(M+H)^+$ intensity increases by a factor of 3 compared to the original peptide.

3.3 Detection Limits

Figure 4 shows a spectrum of Gramicidin D obtained from a total amount of sample of 10^{-15} mol demonstrating the high sensitivity of TOF-SIMS for the detection of peptides in the mass range of 2000 amu [7]. The experimental detection limit confirms the value of $5 \cdot 10^{-15}$ mol minimum sample size calculated from the transformation probability of Gramicidin ($P = 5 \cdot 10^{-5}$).

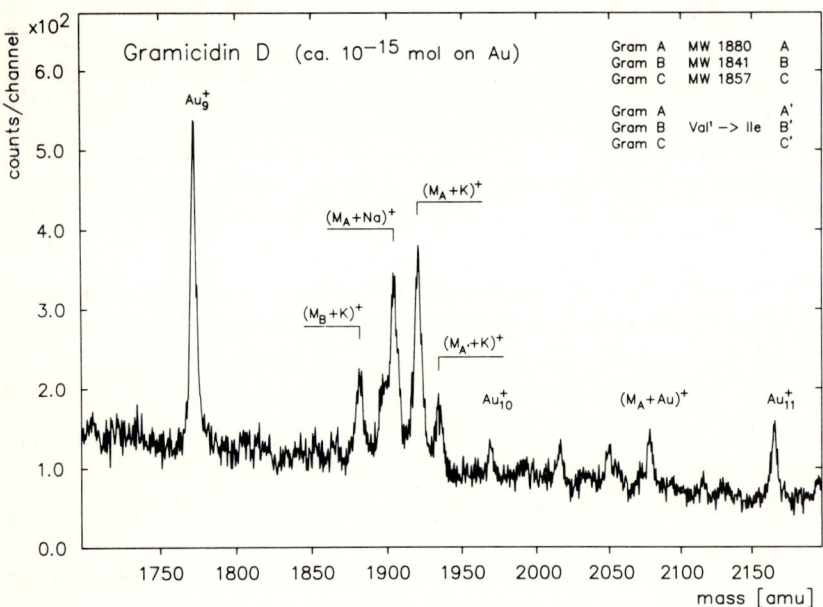

Fig. 4 Positive SIMS spectrum of Gramicidin D; sample: 10^{-15} mol on Au; primary ions: $2 \cdot 10^9$, Ar^+, 12 keV

4. Conclusion

TOF-SIMS turns out to be a sensitive technique for the detection of peptides down to the fmol-range. In combination with HPLC-separation, a quick and reliable identification of tryptic peptides of apolipoproteins is possible. Sequence information can be obtained by TOF-SIMS monitoring of chemical degradation products of analyzed peptides.

A more detailed knowledge about the ion formation process may be obtained by appropriate derivatization experiments. First studies suggest that protonation occurs at basic amino groups whereas cationization is independent of reactive groups within the molecule. The number of NH_2-groups in a peptide has a great influence on absolute as well as relative molecular ion yields, as they are obtained from Ag supported samples.

References

1. P.Steffens, E.Niehuis, T.Friese, A.Benninghoven, in: IFOS II, Springer Series in Chemical Physics, Vol. 25 (1983)
2. P.Steffens, E.Niehuis, T.Friese, D.Greifendorf, A.Benninghoven, J.Vac.Sci.Technol. A3 (3), 1985
3. I.V.Bletsos, D.M.Hercules, D.van Leyen, A.Benninghoven, these proceedings
4. H.U.Jabs, G.Assmann, J.Lipid Res., to be published
5. M.Junack, A.Benninghoven, these proceedings
6. A.Benninghoven, V.Anders, Org.Mass Spectrom. 19, 7 (1984)
7. A.Benninghoven, E.Niehuis, T.Friese, D.Greifendorf, P.Steffens, Org.Mass Spectrom. 19, 7 (1984)

TOF-SIMS of Polymers in the High Mass Range

I.V. Bletsos[1], *D.M. Hercules*[1], *D. van Leyen*[2], *E. Niehuis*[2], and *A. Benninghoven*[2]

[1]Department of Chemistry, University of Pittsburgh, Pittsburgh, PA 15260, USA
[2]Physikalisches Institut der Universität Münster, Domagkstr. 75 D-4400 Münster, F.R.G.

Introduction

First attempts to characterize synthetic polymers by Mass Spectrometry (MS), mainly included pyrolysis [1]. Due to severe degradation processes only pyrolyzates in low mass range could be detected. Invention of softer volatilization and ionization sources like field desorption MS, (FDMS), electrohydrodynamic MS, (EHDMS), [2], ^{252}Cf-plasma desorption (^{252}Cf-PDMS), laser desorption, (LDMS), and secondary ion MS, (SIMS), has expanded MS to include polymer characterization in higher mass ranges, oligomer distributions and surface analysis. FDMS and ^{252}Cf-PDMS have been used to characterize polymers in the high mass range, but limitations in each technique impose restrictions [3,4,5,6]. In FDMS polymer decomposition is temperature dependent and sample preparation and control of ionization process difficult. ^{252}Cf-PDMS requires long data acquisition times. LDMS and SIMS due to severity of ionization process or type of mass analyzer used have been restricted to low mass ranges [7,8,9].

As part of our program to characterize polymers by MS we succeeded in detecting high mass fragments of aliphatic polyamides (nylons) using a time-of-flight secondary ion mass spectrometer (TOF-SIMS) [10]. In general, at low mass range fragmentation of polyamide backbone produces carbon cluster ions and small fragments containing C, H, N and O usually cationized with Ag^+. Protonation and cationization of segments of nylon chains with Ag^+ and Na^+ give rise to a series of quasi-molecular ions of the type $(nR+H)^+$, $(nR+Ag)^+$ and $(nR+Na)^+$ in high mass range, where R is the repeat unit. These peaks are important for identification of the polymer type, and can be used to establish the sequence of repeat units in polymer chains. For nylons of the type -AABB-, e.g. N.66, N.69 etc., cleavage occurs at every other amide linkage consistently throughout the detectable mass range; i.e. the spacing between peaks of quasi-molecular ions corresponds to the repeat unit rather than the individual monomers. In the case where a hydrogen α- to the

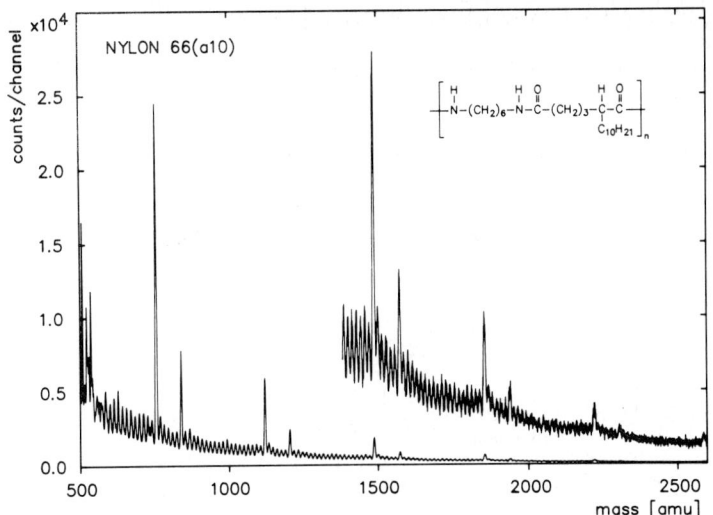

Fig. 1 Positive TOF-SIMS spectrum of Nylon 66(a10)

carbonyl has been substituted by an n-alkyl chain, similar cationization occurs and the pendant group remains intact to the backbone (Fig. 1)

In this report we present mass spectra of a polydimethylsiloxane (PDMS) and some polystyrenes (PS), obtained by a TOF-SIMS, equipped with a mass selected pulsed primary ion source, an angle and time focusing time-of-flight analyzer and a single ion counting detector [11]. Fragmentation at low mass range provided structural information about the repeat unit. Ag^+ cationization of polymer fragments containing a large number of repeat units at high m/z allowed identification of the polymer. Fragmentation patterns were unique for polymers having structurally different repeat units but of equal mass, and identification of such polymers was possible. Oligomer distributions obtained from mass spectra compared well with distributions determined by other techniques for the same polymers.

Results and discussion

A mass spectrum of PDMS is shown in Fig. 2. At m/z < 500, extensive fragmentation is observed producing structurally significant fragments useful to characterize the polymer. At high mass range from m/z = 500 to 10000, intact polymer molecules are detected as cationized units with Ag^+. The most intense peaks at high mass range correspond to $(nR+CH_3+Ag)^+$ series. Presence of a terminal $-CH_3$ in ions detected indicates that desorption of intact polymer molecules occurs without fragmentation. The intensity variation of cationized polymer molecules as a function of mass reflects a distribution of the order of polymeri-

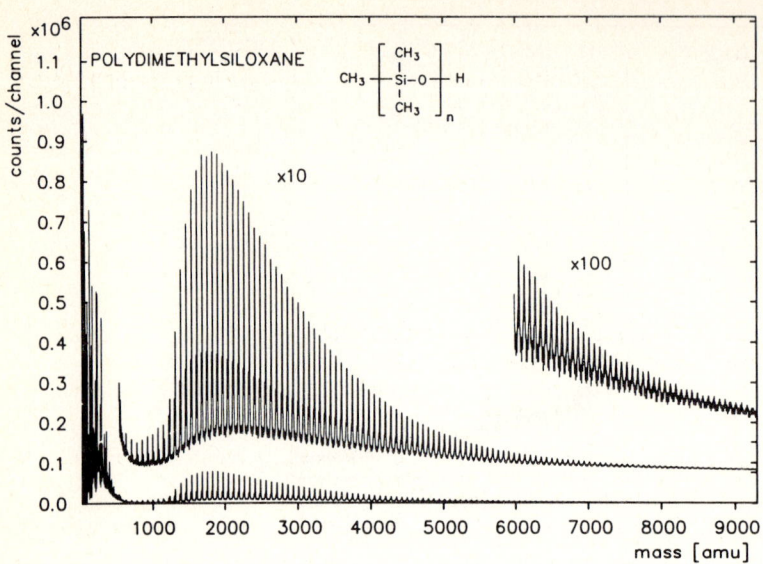

Fig. 2 Positive TOF-SIMS spectrum of Polydimethylsiloxane

zation of PDMS. Loss of terminal $-CH_3$ groups from polymer molecules gives rise to $(nR+Ag)^+$ series of peaks of lower intensity than $(nR+CH_3+Ag)^+$. The largest ion detected at approximately m/z=9600 is cationized with Ag^+, contains a terminal $-CH_3$ group and consists of 128 repeat units of PDMS.

Several polystyrenes with various substituent groups at different positions on the benzene ring or the hydrocarbon backbone were studied. A 1 μl solution of a styrene polymer in toluene (ca. 1×10^{-2} M) was deposited on 100 mm^2 of an etched Ag substrate. A target area of approximately 1 mm^2 bombarded by 12 Kev Ar^+ with an average current of 30 pA, corresponding to a primary ion current density of ~ 10^{-9} A/cm^2 (static SIMS). Part of a poly(α-methylstyrene), P(α-MS), spectrum is shown in Fig. 3. Polymer fragments cationized with Ag^+, $(nR+Ag)^+$ produce most intense peaks; the spacing between them equals one repeat unit. Fragmentation within one repeat unit spacing is consistent throughout the spectrum and is characteristic of the PS type studied. For example the repeat units of P(α-MS) and P(4-MS) have different chemical structures but equal mass (R=118). The most prominent peaks due to Ag^+ cationized fragments for both P(α-MS) and P(4-MS) appear at exactly the same m/z. Fragmentation patterns, however, are different and P(α-MS) and P(4-MS) can be distinguished. Differences in fragmentation are due to different positions of the $-CH_3$ substituent group. Cleavage occurs statistically at each possible bond and peak intensities depend on chemical stability

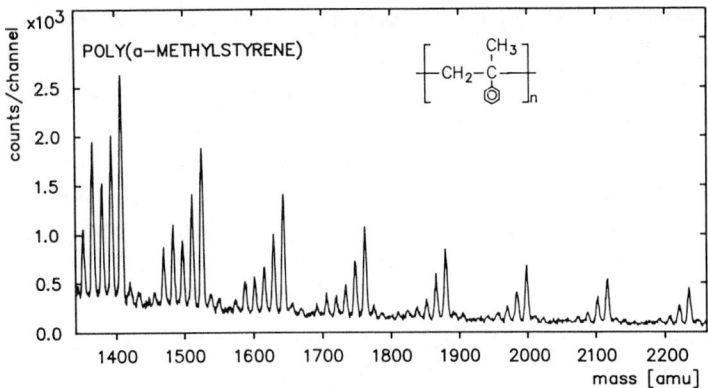

Fig. 3 Positive TOF-SIMS spectrum of Poly(α-Methylstyrene)

of the fragments produced. Detailed discussion on such fragmentation patterns will be published.

Polymer molecular weight distributions obtained from mass spectra were evaluated with polymer standards of known molecular weight distributions determined by gel permeation chromatography (GPC), (e.g. polyethylene, polybutadiene, polystyrene etc.). A spectrum of a PS standard with a number average molecular weight Mn=2964 is shown in Fig. 4 Up to m/z=1800 most intense peaks are due to cationized polymer fragments corresponding to $(nR+Ag)^+$ series. In the range from m/z=1800 to 4500 whole polymer molecules are detected intact with their terminal groups giving rise to $(nR+C_4H_9+Ag)^+$ series of peaks. These peaks are the most intense in this range and their intensity distribution reflects a number of average molecular weight distribution for the PS standard. Loss of terminal $-C_4H_9$ groups results into $(nR+Ag)^+$ series of lower intensity than $(nR-C_4H_9+Ag)^+$. This indicates that the dominant

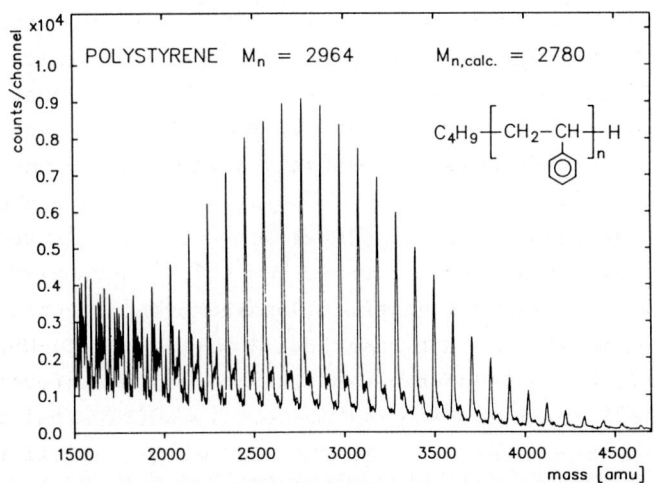

Fig. 4 Positive TOF-SIMS spectrum of Polystyrene

process in the distribution range is the desorption of intact polymer molecules cationized with Ag^+. At the low m/z end of the distribution region, polymer fragments containing a $-C_4H_9$ and whole polymer molecules intact with their terminal groups give rise to peaks which could not be resolved with the instrument used. Consequently, the starting point of the distribution cannot be exactly determined. An Mn, calc.= 2780 was calculated from peak intensities without isotopic contribution corrections. Mn, calc.=2780 is within 6% of the given Mn=2964. A lower than expected Mn, calc. could be due to decreasing detection efficiency, as m/z of ions increases, or to statistically preferential fragmentation of longer chains than shorter ones. For determination of more accurate molecular weight distributions, especially at high m/z ranges, appropriate corrections for detection efficiencies could be necessary.

In conclusion, it has been shown that TOF-SIMS can be used effectively to characterize polymers in low and high mass range and obtain oligomer distributions.

This work was supported by the National Science Foundation under Grant CHE8411835 and by the Deutsche Forschungsgemeinschaft. We are grateful to the Alexander von Humboldt Foundation for providing a senior fellowship for D.M.H., which stimulated this work.

References

1   Y.Sugimura, T.Nagaya, S.Tsunge, Macromolecules, 14, 520 (1981)
2   K.W.S.Chan, K.D.Cook, Org.Mass Spectrom., 18, 423 (1983)
3   T.Matsuo, H.Matsuda, I.Katakuse, Anal.Chem., 51, 1329 (1979)
4   R.P.Lattimer, H.-R.Schulten, Int.J.Mass Spectrom. Ion Phys., 51, 105 (1983)
5   U.Bahr, I.Luderwald, R.Muller, H.-R.Schulten, Angew.Makromol. Chem., 120, 163 (1984)
6   B.T.Chait, J.Shpungin, F.H.Field, Int.J.Mass Spectrom. Ion Process, 58, 121 (1984)
7   J.A.Gardella, S.W.Graham, D.M.Hercules, in "Polymer Characterization"; C.D.Craver, Ed., Advances in Chemistry Series 203, ACS; Washington, D.C. 1983
8   D.Briggs, Surf.Interface Anal., 4, 151 (1982)
9   D.E.Mattern, D.M.Hercules, in press
10 I.V.Bletsos, D.M.Hercules, D.Greifendorf, A.Benninghoven, Anal.Chem., 57, 2384 (1985)
11 P.Steffens, E.Niehuis, D.Greifendorf, A.Benninghoven, J.Vac.Sci. Technol., 3, 1322 (1985)

The Application of Time-of-Flight Secondary Ion Mass Spectrometry in the Characterization of Apolipoprotein Mutants

H.-U. Jabs[1], M. Walter[1], G. Assmann[1], and A. Benninghoven[2]

[1]Institut für Klinische Chemie und Laboratoriumsmedizin,
Westfälische Wilhelms Universität Münster,
Albert-Schweitzer-Strasse 33, D-4400 Münster, F. R. G.
[2]Physikalisches Institut der Universität Münster,
D-4400 Münster, F. R. G.

Lipoproteins are complexes composed of apolipoproteins and non-covalently bound lipids which are found in human plasma. They can be isolated from plasma by sequential ultracentrifugation and they are named according to their floating densities: very-low-density lipoproteins (VLDL), low-density lipoproteins (LDL) and high-density lipoproteins (HDL). They differ in lipid to protein ratios and in their apolipoprotein composition. In epidemiological studies it was shown that high concentrations of LDL in plasma are associated with a high risk for coronary heart disease, whereas a high concentration of HDL seems to be a protection factor for a myocardial infarction. Apolipoproteins are separated from lipoproteins by the extraction of lipids with ethanol and ether. Apolipoproteins bind lipids, serve as cofactors of lipolytic enzymes and represent ligands for the cellular recognition of lipoproteins on plasma membrane-bound receptors[1,2]. Major structural components of VLDL are apolipoproteins B, E, C-II and C-III, whereas the major apolipoprotein of LDL is apo B, while HDL is primarily composed of apolipoproteins A-I and A-II. The amino acid sequence of all apolipoproteins is known [3]. Apo A-I is a cofactor in the activation of the cholesterol esterifying enzyme lecithin-cholesterol acyltransferase (LCAT) and seems to be the specific ligand in the cellular uptake of HDL [4].

In the past, several structural variants of apo A-I have been detected in screening analyses of human plasma by isoelectric focusing procedures [5-7]. These variants differ in charge due to their amino acid substitutions at various sites of the molecule. The discovery of apolipoprotein mutants is of great interest for the further understanding of the metabolic functions of apolipoproteins. For the structural characterization of apolipoprotein mutants a combination of isoelectric focusing, high performance liquid chromatography and time-of-flight secondary ion mass spectrometry (TOF-SIMS) was used.

Lipoproteins were isolated from human plasma by ultracentrifugation and were subsequently delipidated by the extraction of lipids with ethanol and ether. The obtained apolipoproteins were purified using isoelectric focusing in immobilized pH-gradients (Immobiline, LKB Bromma). This technique allowed focusing in very narrow pH-gradients of 0.5 pH units per 10 cm. The purified apolipoproteins were electro-eluted from the gels and were subsequently digested with trypsin. Tryptic peptides were separated by reverse phase HPLC chromatography. A characteristic HPLC chromatogram is presented in Fig. 1. Peptides in the HPLC fractions were identified by manual amino acid sequencing using the DABITC method [8] and by determination of molecular masses using TOF-SIMS mass analysis [9-11]. The position of the tryptic peptides in the primary structure of apolipoprotein A-I is indicated in table 1. Representative mass spectra of four different tryptic apo A-I peptides are shown in Fig. 2.

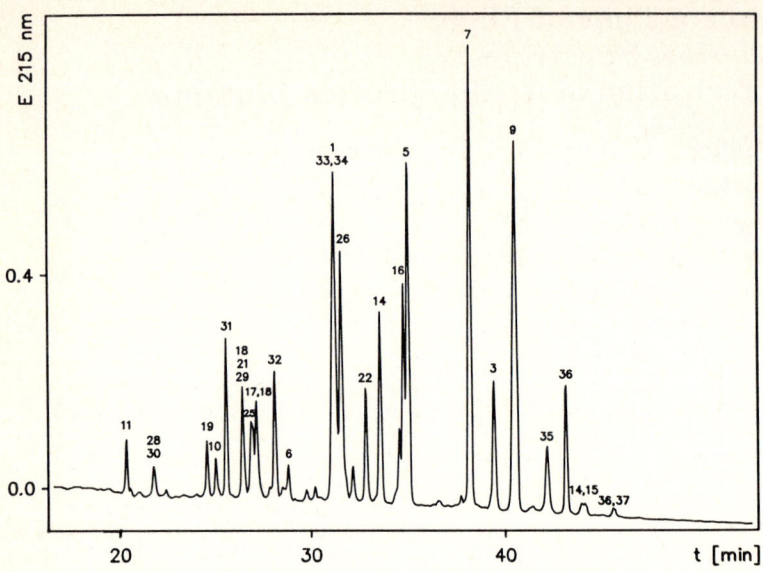

Fig. 1. HPLC chromatogram of tryptic peptides from apolipoprotein A-I Numbers of peptides are indicated

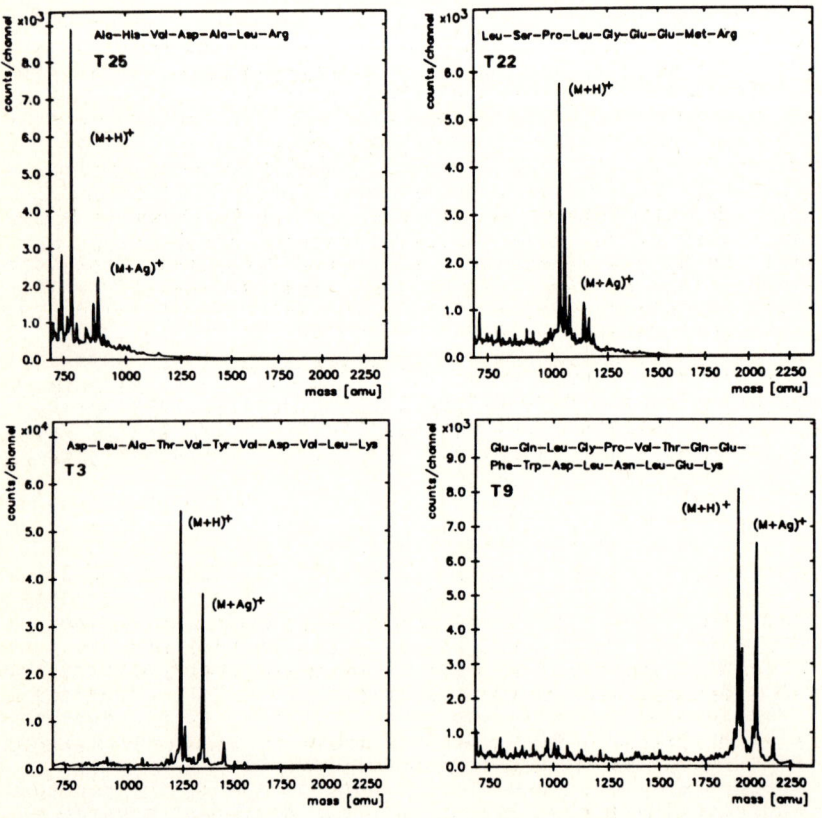

Fig. 2. TOF-SIMS mass spectra of four tryptic apo A-I peptides

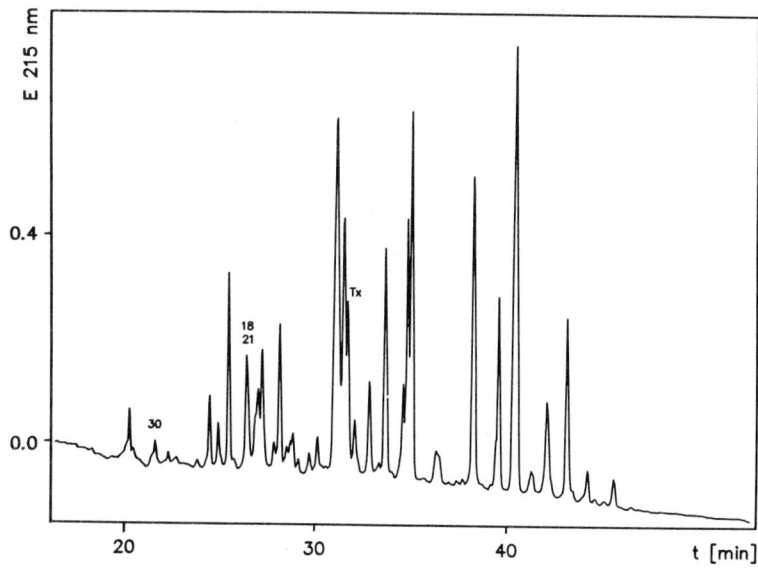

Fig. 3. HPLC chromatogram of tryptic peptides of the mutant apo A-I
The newly occurring peptide is indicated as Tx

In a screening program using analytical isoelectric focusing in immobilized pH-gradients, we found a mutant apo A-I (Münster-1) which differed in pI from normal apo A-I by only 0.01 pH units. The familial nature of this mutant protein was established by pedigree analysis. The high resolution power of the Immobiline gels allowed us to separate the mutant from the normal apolipoprotein. After tryptic digestion peptides of both proteins were separated by HPLC chromatography. The HPLC chromatogram of the mutant apo A-I is shown in Fig.3. The HPLC chromatogram of the mutant apolipoprotein differs from the control preparation by the appearance of a newly generated peptide (Tx) with a molecular weight of 964 D and by a decrease in the height of the peaks normally containing peptides (T28, T30) and (T18, T21, T29), respectively. In these peaks the peptides T28 and T29 could not be detected. The mass spectrum of the newly occurring peak is shown in Fig. 4. With manual micro-sequencing the following five amino acids of peptide Tx were found: Leu-Ala-Ala-His-Leu. Assuming a point mutation, the molecular weight and the sequence analysis of peptide Tx can only be explained by a substitution of arginine in position 177 of the amino acid sequence by histidine.

T_{28}(429 D) Leu-Ala-Ala-Arg$_{177}$ Leu-Glu-Ala-Leu-Lys T_{29}(572 D)

T_x (964 D) Leu-Ala-Ala-His-Leu-Glu-Ala-Leu-Lys

The combined techniques of isoelectric focusing, reverse phase HPLC separation of proteolytic peptides and subsequent molecular mass analysis by TOF-SIMS constitute a new tool in the analysis of proteins and the detection of protein mutants.

With these techniques a mutant apolipoprotein is characterized step by step. First, mutant apolipoproteins were found in a screening program using isoelectric focusing. Second, the location of the mutation in the amino acid sequence could be restricted to a tryptic peptide by HPLC chromatography. Third, with TOF-SIMS mass analysis

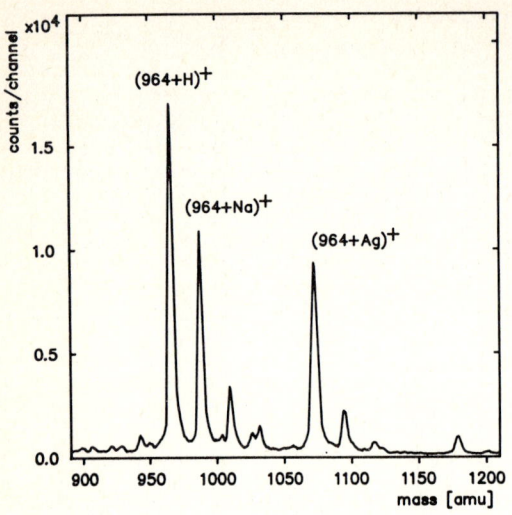

Fig. 4. TOF-SIMS mass spectrum of tryptic peptide Tx from mutant apolipoprotein A-I (Münster-1)

Table 1. Amino acid sequence of human apolipoprotein A-I [3] Tryptic peptides and molecular masses as determined by TOF-SIMS are indicated

Sequence	Peptide (Mass)
Asp-Glu-Pro-Pro-Gln-Ser-Pro-Trp-Asp-Arg-	T- 1 (1226)
Val-Lys-	T- 2
Asp-Leu-Ala-Thr-Val-Tyr-Val-Asp-Val-Leu-Lys-	T- 3 (1235)
Asp-Ser-Gly-Arg-	T- 4 (433)
Asp-Tyr-Val-Ser-Gln-Phe-Glu-Gly-Ser-Ala-Leu-Gly-Lys-	T- 5 (1400)
Gln-Leu-Asn-Leu-Lys-	T- 6 (614)
Leu-Leu-Asp-Asn-Trp-Asp-Ser-Val-Thr-Ser-Thr-Phe-Ser-Lys-	T- 7 (1612)
Leu-Arg-	T- 8
Glu-Gln-Leu-Gly-Pro-Val-Thr-Gln-Glu-Phe-Trp-Asp-Asn-Leu-Glu-Lys-	T- 9 (1932)
Glu-Thr-Glu-Gly-Leu-Arg-	T-10 (703)
Gln-Glu-Met-Ser-Lys-	T-11 (621)
Asp-Leu-Glu-Glu-Val-Lys-	T-12 (731)
Ala-Lys-	T-13
Val-Gln-Pro-Tyr-Leu-Asp-Asp-Phe-Gln-Lys-	T-14 (1252)
Lys-	T-15
Trp-Gln-Glu-Glu-Met-Glu-Leu-Tyr-Arg-	T-16 (1283)
Gln-Lys-	T-17
Val-Glu-Pro-Leu-Arg-	T-18 (612)
Ala-Glu-Leu-Gln-Glu-Gly-Ala-Arg-	T-19 (872)
Gln-Lys-	T-20
Leu-His-Glu-Leu-Gln-Glu-Lys-	T-21 (895)
Leu-Ser-Pro-Leu-Gly-Glu-Glu-Met-Arg-	T-22 (1031)
Asp-Arg-	T-23
Ala-Arg-	T-24
Ala-His-Val-Asp-Ala-Leu-Arg-	T-25 (780)
Thr-His-Leu-Ala-Pro-Tyr-Ser-Asp-Glu-Leu-Arg-	T-26 (1301)
Gln-Arg-	T-27
Leu-Ala-Ala-Arg-	T-28 (429)
Leu-Glu-Ala-Leu-Lys-	T-29 (572)
Glu-Asn-Gly-Gly-Ala-Arg-	T-30 (602)
Leu-Ala-Glu-Tyr-His-Ala-Lys-	T-31 (830)
Ala-Thr-Glu-His-Leu-Ser-Thr-Leu-Ser-Glu-Lys-	T-32 (1215)
Ala-Lys-	T-33
Pro-Ala-Leu-Glu-Asp-Leu-Arg-	T-34 (812)
Gln-Gly-Leu-Leu-Pro-Val-Leu-Glu-Ser-Phe-Lys-	T-35 (1230)
Val-Ser-Phe-Leu-Ser-Ala-Leu-Glu-Glu-Tyr-Thr-Lys-	T-36 (1386)
Lys-	T-37
Leu-Asn-Thr-Gln-	T-38 (474)

the amino acid exchange could be determined by comparing the masses of new peptides with the masses of disappearing peptides.

Thus, with the described techniques it is feasible to monitor for genetic polymorphisms and translational modifications on a nano-molar scale. Depending on the structure of the peptide or glycopeptide investigated, conclusions on the location of the mutation or modification can be drawn from the HPLC chromatogram in conjunction with a molecular mass analysis. In some cases it may, however, be necessary to confirm the anticipated structural change by the use of either different proteolytic enzymes or additional amino acid sequencing. In these instances the combination of HPLC chromatography and TOF-SIMS mass analysis still is of great advantage in the preselection of the right peptide for further studies.

References

1. P.N. Herbert, G. Assmann, A.M. Gotto,Jr, and D.S. Fredrickson: in The Metabolic Basis of Inherited Disease. 1982. Editors: J.B. Stanbury, J.B. Wyngaarden, D.S. Fredrickson, J.L. Goldstein and M.S. Brown, McGraw-Hill, New York, pp 589-621.
2. G. Assmann: Lipid Metabolism and Atherosclerosis (Schattauer-Verlag, Stuttgart, New York, 1983)
3. H.B. Brewer,Jr, T. Fairwell, A. Larne, R.Ronan, A. Houser and T.J. Bronzert: Biophys.Res.Commun.$\underline{80}$, 623 (1978)
4. G. Schmitz, H. Robenek, U. Lohmann and G. Assmann: EMBO $\underline{4}$, 613 (1985)
5. H.J. Menzel, R.G. Kladetzky and G. Assmann: J.Lipid Res. $\underline{23}$,915 (1982)
6. H.J. Menzel, G. Assmann, S.C. Rall et al.:J. Biol. Chem. $\underline{259}$, 3070, (1984)
7. G. Assmann, H.J. Menzel, R.G. Kladetzky and G. Büttner: J. Clin. Chem. Clin. Biochem. $\underline{22}$,585, (1984)
8. J.Y. Chang, D. Brauer, B. Wittmann-Liebold: FEBS Lett. $\underline{93}$, 205 (1978)
9. P. Steffens, E. Niehuis, T. Friese and A. Benninghoven: in Springer series: Chemical Physics, Vol 25, 1983, pp 111-117
10. A. Benninghoven and V. Anders: Org. Mass Spectrom. $\underline{19}$, 345, (1984)
11. A. Benninghoven and W. Sichtermann: Anal. Chem. $\underline{50}$, 1180, (1978)

Part III

Liquid SIMS Including FAB

Sputtering Yields from Liquid Organic Matrices

D.F. Barofsky and E. Barofsky

Department of Agricultural Chemistry, Oregon State University, Corvallis, OR 97331, USA

1. Introduction

Accurate measurements of secondary particle yields should provide significant insights into the mechanisms of secondary ion emission from organic liquids. Variations in primary beam energy, incident flux density, and primary particle mass, for example, are all manifested in changes in total particle yield. Ratios of secondary ion yields from two different species can be sensitive, quantitative monitors of the chemistry and kinetics respectively of ionization processes.

Previously, we reported on a straightforward gravimetric method for estimating total particle yields from liquid, organic substrates [1]. We have improved on this procedure, and in this paper we present preliminary results of its use to estimate total particle yields from polyethylene glycol 400 (PEG 400).

2. Experimental

Primary particles were generated with a liquid metal ion gun [2]; Ga, In, and Si-Au emitters were employed. Two gun configurations were used; neither had a focusing capability. The first, which has been described elsewhere [2], has no provision for varying the gun's operating voltage and the energy of the ion beam independently. The second, shown schematically in Fig. 1, permits independent variation of these two parameters, but its energy can only be varied from 5 to 8 keV due to a focusing

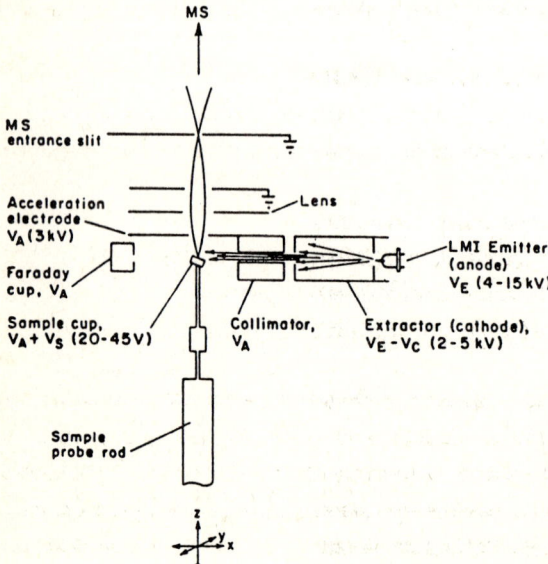

Fig. 1 Schematic of a liquid metal ion primary gun which permits independent variation of the gun's operating voltage and the ion beam's energy.

effect which occurs between the extractor and collimator. Current densities for the primary beams ranged from 3×10^{-7} A/cm^2 to 3×10^{-6} A/cm^2. Sample irradiation times were varied from 20 to 60 min.

Only pure glycerol and pure PEG 400 were used as samples in this study. The evaporation rate for pure glycerol was determined previously to be 0.30 mg/cm$^2 \cdot$min [1]; the evaporation rate for pure PEG 400 was determined in the present study to be 0.88 µg/cm$^2 \cdot$min.

Two types of data sets were acquired. The first was obtained from several trials in which the irradiation dose was maintained constant from trial to trial, and the results of all the trials were averaged. The second resulted from several trials in which the irradiation dose was varied either by changing the emission current of the liquid metal ion gun or by varying the irradiation times, and the results were interpreted by regression analysis, as will be explained below.

Determinations of total yields based on gravimetric measurements are subject to gross, systematic errors arising from evaporation losses, exchanges of volatile materials (e.g. H_2O and MeOH) with the atmosphere, ill-defined or misjudged equilibrium states used as standard states for taking readings, and inaccuracies in calibration. If the effects of these systematic factors are summed, the total loss of mass of sample material (L) can be represented by the expression

$$L = \Sigma(\text{Systematic Factors}) + \text{Yield} \times D \tag{1}$$

where D is the dose. In the constant-dose method an estimate of the magnitudes of the systematic factors is attempted and the sputtering component of the mass-loss (L_s) is calculated from the relation

$$L_s = L - \Sigma(\text{Systematic Factors}). \tag{2}$$

The yield is then computed from its definition

$$\text{Yield} = L_s/D. \tag{3}$$

Preliminary estimates of total particle yields from pure glycerol, bombarded by 3-5 keV indium ions, were determined by this technique to be in the range of 1000-3000, two to three orders of magnitude higher than those generally observed for metals in this energy range [1].

In the variable-dose method the yield is obtained directly, without estimating the magnitudes of the systematic factors, as the slope of a linear regression, $L_{reg}(D)$, performed on a set of loss-dose data:

$$\text{Yield} = dL_{reg}/dD. \tag{4}$$

This differential approach is adapted from procedures which have long been used in measurements of sputtering yields from solids [3].

It is clear from (1) that the yield will be the derivative of sample-loss with respect to dose providing both the systematic factors and the yield are independent of dose. For samples with relatively high evaporation rates, such as glycerol, the first assumption will not hold if the dose is varied by changing exposure times, since the amount of loss suffered by the sample through evaporation would also be affected in the process. This is not the case however with PEG 400 which has a very low evaporation rate. The assumption regarding yield becomes invalid if the current density of the primary ion beam exceeds about 10^{-5} A/cm^2; beyond this rate of dose the sample's rate of loss tends in the limit to saturate, i.e. to become independent of the primary beam's current [4].

3. Results

Loss-dose data for PEG 400 bombarded with In-ions at two different energies is plotted in Fig. 2; total yields estimated from this data by means of the variable-dose method are 630 and 850 respectively for 5.3 keV and 6.0 keV In-ions. The coefficients of determination, R^2, adjusted for the number of degrees of freedom is 99.6% for both of the regression line shown in Fig. 2. All other experiments conducted to date at these same energies resulted in values for R greater than 90% and values for the yields within one standard deviation of those obtained in Fig. 2.

The average total yields from PEG 400 for three different primary ions are plotted in Fig. 3 as a function of the mass of the primary particles. The yields were determined from constant-dose measurements. The regression curve for yield versus mass is also drawn in the figure.

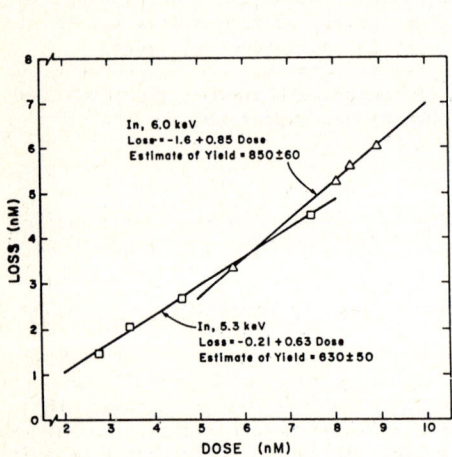 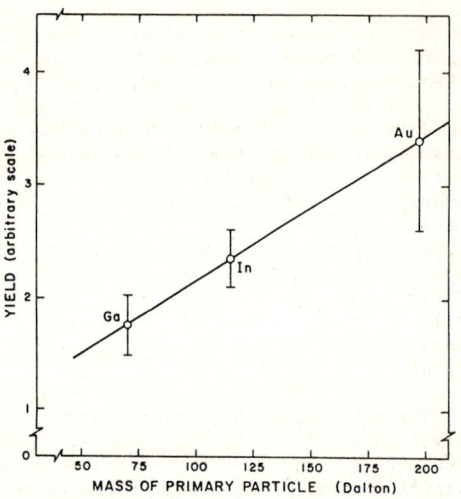

Fig. 2 Loss-dose data for PEG 400 bombarded with In-ions at two different energies.

Fig. 3 Total yield of PEG 400 versus mass of primary particle. The energy of all three primary particle beams was 8 keV.

4. Discussion

The yields for PEG 400 are plotted in Fig. 4 as a function of energy along with the data previously obtained for glyercol [1]. The PEG 400 values, while large on an absolute scale, are considerably lower than those for glycerol. There may be fundamental explanations for this difference, but we believe the primary reason for most of it is inaccuracy in the glycerol values. This supposition will be tested by extending the range of energies for PEG 400 and redetermining the glycerol yields using the variable-dose method.

The very good fit of a straight line to the data in Fig. 3 can only be regarded as fortuitous. The size of the error associated with the individual yield values precludes any quantitative conclusions about a model for this data. More precise determinations of yields using the variable-dose technique may be capable of providing some insight into the nature of the yield's dependence on the masses of the primary particles.

Our total yield data is still fragmentary, and we have yet to initiate measurements of secondary ion yields. Nonetheless a picture of some aspects of the nature of

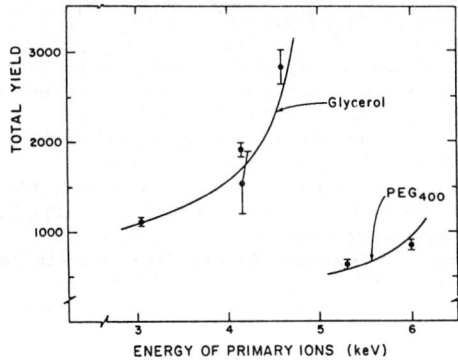

Fig. 4 Total yield versus energy of primary ion for pure glycerol and pure PEG 400.

sputtering from organic liquids seems to be emerging, and it is worth commenting on these features at this juncture.

Intuitively, it is difficult to conceive of a mechanism whereby several hundreds to possibly thousands of secondary molecules are all removed from a single surface layer during an individual impact event. For example, in the case of glycerol (with an average molecular diameter of about 0.5 nm) this would require that the majority of ejected molecules come from a region 15-25 molecular diameters away from the impact point. Experimentally, typical damage cross-sections for metals have radii of about 0.5-0.6 nm; this is equivalent to only three or four atomic diameters. Monte Carlo simulation of the trajectories of recoiled copper atoms following the impact of a 10 keV argon atom shows that only a very few particles can be expected to eject into the vacuum from as far away as 3-4 nm (15-20 atomic diameters) from the point of impact of the primary particle [5].

The more plausible interpretation for the large magnitudes of the total yields from glycerol and PEG 400 would be that the majority of molecules ejected during each individual impact event arise from the bulk of the liquid matrix along or in the near vicinity of the trajectory of the primary particle. It seems likely that a majority of the radiation-damaged molecules, which would be close to the primary particle's trajectory, would be among the vaporized particles, that is each impact event would automatically leave the sample matrix purged of most of the debris from that event. If pre-existing ions are homogeneously distributed in the matrix, presumably a proportionate number of them would also accompany the ejected material. Essentially the same picture has been proposed by Wong et al. [6,7]. Their proposed model, which is supported not only by yield determinations but also by ion abundance measurements as a function of time, also addresses the question of surface supply mechanisms and surface renewal. They conclude from their time dependent data that the products of radiation damage are, in fact, entirely removed from the liquid surface by the sputtering process, and that restoration of the surface by diffusion of molecules to or from the bulk is thus obviated.

These ideas concerning sputtering from organic liquids are consistent with the frequently reported requirements of solubility of the sample compound and of homogeneity of the sample solution as prerequisites to producing mass spectra [8,9] and with the growing number of mass spectrometric observations which suggest that nonradical molecular ions are preformed in the sample solution on the end of the insertion probe.

Given the total yields we observe and the range of current densities or equivalent current densities (10^{-6}-10^{-5} A/cm^2) normally employed in liquid-matrix-assisted SIMS, depletion of the liquid surface must be occurring at rates of 1-15 ng/cm$^2 \cdot$s. In the case of homogeneous liquids with no surface active components Wong et al. contend that there should be no surface damage in the sense of irradiated solids, but simply the continuous creation of new surface with each impact event [6,7]. However in the case of surfaces with active species, that is species enriched or depleted in the surface layer, there must be some form of surface disruption due to particle bombardment. Using a diffusion model, Magee [10] calculated that for the effects of such a disruption to register in the sputtered products, the surface would have to be depleted by sputtering at a rate greater than 3×10^{16} particles/$cm^2 \cdot s$, i.e. more than 4 μg/$cm^2 \cdot s$ for glycerol. Whichever the case, it seems very likely that under the usual conditions for liquid-matrix-assisted SIMS each primary particle impacts on an undisturbed, intact surface. This is a low dosage condition and implies a role for surface activity in the process of sputtering from organic liquids, a feature which might (at least partially) account for past observations [11,12] in which certain, surface-active species seemed to promote removal of matrix material and consequently ion emission.

5. Acknowledgement

This work was supported by the Department of Health and Human Services, Public Health Service, NIADDK Grant R01 AM 36050.

6. References

1. D.F. Barofsky, A.M. Ilias, E. Barofsky, and J.H. Murphy, Proceedings of the 32nd Annual Conference on Mass Spectrometry and Allied Topics (San Antonio 1984), A.S.M.S., p. 182.

2. D.F. Barofsky, U. Giessmann, A.E. Bell, and L.W. Swanson, Anal. Chem. 55, 1318 (1983).

3. O. Almen and G. Bruce, Nucl. Instrum. Methods 11, 279 (1961).

4. D.F. Barofsky, U. Giessmann, L.W. Swanson, and A.E. Bell, Proceedings of the 29th International Field Emission Symposium (Göteborg 1982), H-O. Andrén and H. Nordén, Eds., Almqvist and Wiksell Int., Stockholm, 1982, p. 425.

5. T. Ishitani and R. Shimizu, Phys. Lett. 46A, 487 (1974).

6. S.S. Wong, F.W. Röllgen. I Manz, and M. Przybylski, Biomed. Mass Spectrom. 12, 43 (1985).

7. S.S. Wong, K.P. Wirth, and F.W. Röllgen, in Springer Series in Chemical Physics - Ion Formation from Organic Solids (IFOS III), these proceedings.

8. S.A. Martin, C.E. Costello, and K. Biemann, Anal. Chem. 54, 2362 (1982).

9. M. Przybylski, Fres. Z. Anal. Chem. 513, 402 (1983).

10. C.W. Magee, Int. J. Mass Spectrom. Ion Phys. 49, 211 (1983).

11. M. Barber, R.S. Bordoli, G.J. Elliott, R.D. Sedgwick, and A.N. Tyler, J. Chem. Soc., Faraday Trans. 79, 1249 (1983).

12. W.V. Ligon and S.B. Dorn, Int. J. Mass Spectrom. Ion Processes 57, 75 (1984).

Sputtering from Liquid and Solid Organic Matrices

S.S. Wong, K.P. Wirth, and F.W. Röllgen

Institute of Physical Chemistry, University of Bonn, Wegelerstr. 12,
D-5300 Bonn, F.R.G.

1. Introduction

The use of a liquid matrix in organic secondary ion mass spectrometry (SIMS) provides several advantages [1]: Firstly, high and long lasting secondary molecular ion signals are obtained from samples by the application of high incident particle fluxes. Secondly, abundant molecular ion signals are recorded from thermally very labile large molecules. Thirdly, knowledge of solvent chemistry, ion chemistry in solution and colloid chemistry can be utilized to control the intensity and type of molecular ions formed, the kind of fragments, and the level of fragmentation, to a significant extent.

We have performed experiments to elucidate the role of the liquid matrix in the sputtering of molecular ions by keV Xe/Xe$^+$ particle impact. In the following results are reported for glycerol, which is most widely used in fast atom bombardment mass spectrometry (FAB MS).

2. Experimental

A modified AEI-MS9 double focusing mass spectrometer equipped with a home-built FAB secondary ion source was used in these investigations. The primary Xe atom beam (about $3 \cdot 10^{13}$ atoms/cm^{2}s [2]) composed of ions and neutrals was generated in a saddle field discharge source operated at 6 kV. The beam diameter was larger than the target area and the angle of incidence was 45°. The mean energy of the Xe/Xe$^+$ atoms was estimated to be about 6 keV at the target surface set on +4 kV. The horizontal position of the target allowed the deposition of thick layers on the probe surface. The layer thickness was typically about 0.5 mm.

3. Sputtering and Ionization Yields

The sputtering yield is the most important quantity needed for discussion of the sputtering mechanism. Recently, we have measured volume sputtering yields of the following liquids [2]: a) pure glycerol: 200 nm^3/incident Xe particle, b) a mixture of stachyose, ammonium chloride and glycerol (ratio of weights 1:1:6): 170 nm^3/particle, and c) a mixture of the pentapeptide, PZ-Pro-Leu-Gly-Pro-Arg, with glycerol (ratio of the weights 1:24): 110 nm^3/particle. The sputtering yields are deduced from volumetric measurements of the erosion rates. The uncertainty of these values is probably not better than a factor of two, mainly due to the inaccuracy in the determination of the incident particle flux.

The above sputtering yield for the stachyose mixture is by a factor of 1.7 higher than that reported previously [3], based on a different method to determine the incident particle flux. For pure glycerol our value (about 1700 glycerol molecules/6 keV Xe atom) is smaller than that communicated previously by others [4], however, recent measurements of the same group point to a correction of their earlier data to a smaller value [5]. It is obvious that more accurate measurements are needed, although sputtering yields of keV particles for organic liquids in the range of several hundred to more than a thousand molecules seem to be established.

A positive secondary ion yield of about 1.5 ions/incident particle was measured for the stachyose mixture. Disregarding a mass mass dependent discrimination of secondary ions by the mass spectrometer, the FAB mass spectrum of the mixture gives a secondary $(M + NH_4)^+$ molecular ion yield of about 10^{-3} molecular ions/incident particle. Combining this result with the above sputtering yield and the mixing ratio an ionization efficiency of about $5 \cdot 10^{-5}$ is derived, equivalent to about 1 Coul/mole. It has to be noted that this ionization efficiency is related to a sputtering rate and not deduced from a complete consumption of the sample layer, which gives a smaller value for the ionization efficiency.

4. Surface Renewal and Surface Supply Mechanisms

The tetrasaccharide stachyose is thermally very labile, non-volatile, and has no surface-active property regarding glycerol as solvent. Accordingly, stachyose was chosen as test compound to study the mechanism of surface renewal during sputtering at high incident particle fluxes. For molecular ion formation by ammonium attachment, ammonium chloride was added to the glycerol solution. Investigations of the molecular ion intensity, the fragmentation level and the chemical noise during prolonged sputtering of thick sample layers revealed that almost no products of radiation damage are left on the surface, provided a sufficient amount of glycerol is still present in the mixture [2,3]. A significant contribution to molecular ion emission by diffusion of molecules from the bulk could be excluded. Therefore it was concluded that the renewal of the surface of the sample layer has to be attributed to the sputtering process itself, removing most of the radiation damage formed by the collision cascade. A prerequisite of this sputtering mechanism is high sputtering yields.

Although a supply of molecules from the bulk to the surface is not needed for maintaining a continuous molecular ion emission under dynamic bombarding conditions, it contributes to the molecular ion formation for surface-active compounds. This has been shown for surfactants [1,6] and has recently also been found for the above pentapeptide [2]. The molecular ion intensity of the pentapeptide decreases much steeper during continuous bombardment of thick layers than the molecular ion signal of saccharides having no surface active properties. From the decrease of the intensity curve and the observation that there is still a liquid layer of glycerol and products of radiation damage left on the probe surface at the time the molecular ion signal of the peptide disappears into the chemical noise, it was concluded that ion migration is involved in the transport of the peptide to the surface [2]. The peptide is basic and protonized by glycerol. Therefore, the peptide is charge carrier in the solution and thus should contribute to the current flow (in the order of 1 $\mu A/cm^2$) from the bulk to the surface to compensate for the charges extracted from the surface during bombardment.

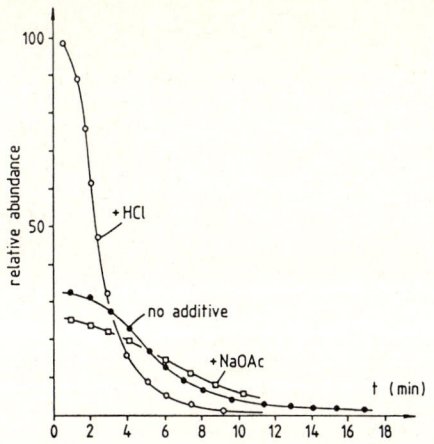

Fig. 1 Dependence of the $(M + H)^+$ molecular ion intensity of PZ-Pro-Leu-Gly-Pro-Arg on sputtering time without and with additives. Weight ratio of the pentapeptide and glycerol 1:20.

We have performed experiments with additives such as HCl and NaOAc to prove the assumption of a surface supply of peptide molecules by ion migration. HCl raises the concentration of protonized peptide molecules involved in the transport to the surface while NaOAc has the opposite effect of decreasing the contribution of peptide molecules to the current flowing to the surface, which is mainly performed by mobile Na^+ ions. Accordingly, as shown in Fig.1, the molecular ion intensity decreases more slowly with NaOAc and more steeply with HCl added to the solution than the intensity curves without additives. Of course, both the additives also change the ionization efficiency for molecular ion formation.

Similar experiments with quaternary compounds also suggest a supply of cations from the bulk of the solution to the surface by ion migration during sputtering. Thus, derivatization by quaternization [7-9] is also in favour of an enhanced surface supply by ion migration.

5. Sputtering Mechanism

The sputtering mechanism has to account for high sputtering yields and the ejection of molecules and molecular ions of thermally labile compounds. A spraying type of sputtering mechanism in which the collision cascade causes a gasification of matrix/sample molecules leads to the ejection of molecular clusters with subsequent fast decomposition. This explains most easily the observed sputtering phenomena [2].

Evidence of the desorption of vibrationally cold molecular ions has been obtained by the observation of abundant $(M + H)^+$ molecular ions of stachyose in the FAB mass spectrum of a mixture of stachyose, tartaric acid and glycerol (Fig.2). The $(M + H)^+$ ion is extremely labile. Under isothermal conditions of field desorption this molecular ion could be only detected as a weak signal between about 50 and 100 °C. It is concluded that the mean vibrational temperature of this ion in the spectrum of Fig.2 is below 100 °C.

6. Sputtering From a Frozen Glycerol Matrix

The advantages of secondary ion formation by sputtering of molecules from a solid ammonium chloride matrix have been demonstrated for static SIMS conditions, using sucrose among others as test compound [10].

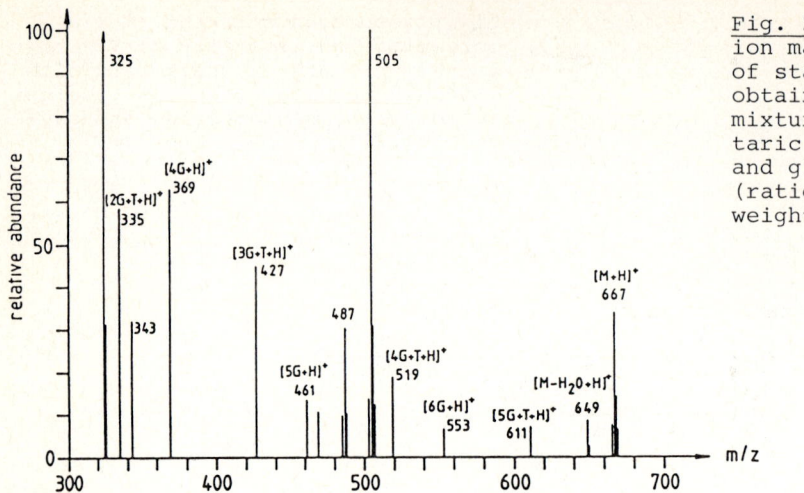

Fig. 2 Secondary ion mass spectrum of stachyose obtained from a mixture with tartaric acid (T) and glycerol (G) (ratio of the weights 1:0.5:10).

However, we were unable to detect molecular ions of sucrose from a solid mixture with ammonium chloride under dynamic bombarding conditions. This result, and the question concerning the difference in sputtering from a liquid and solid matrix, prompted us to perform experiments with a frozen glycerol matrix.

In Fig.3 the FAB spectra of a liquid and frozen mixture of glucose, NH_4Cl and glycerol are compared [2]. In contrast to the liquid, the frozen mixture yields practically no molecular ion of glucose, but an abundant fragment ion which arises from elimination of water. This effect can be attributed to a higher molecular stress exerted to molecules in sputtering from a frozen matrix. The higher stress arises from a larger amount of energy required in the collision cascade for the instantaneous ejection of molecules from the solid state, because melting is not involved in the sputtering process [2]. Therefore, a liquid state of the sample is a prerequisite for soft ionization of thermally labile molecules under dynamic bombarding conditions.

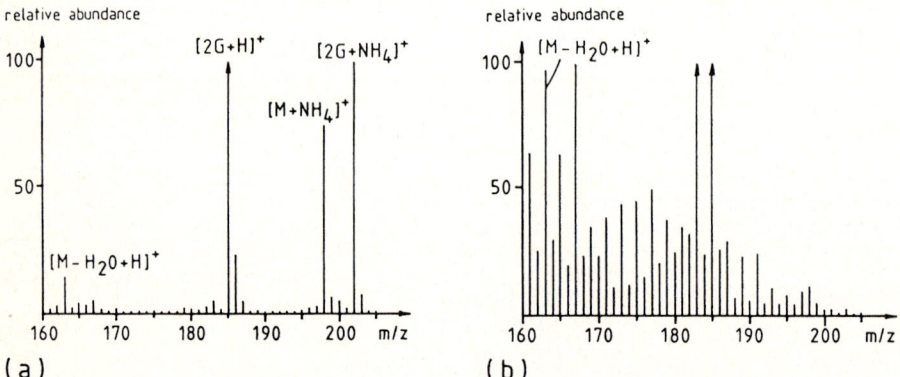

Fig. 3 Secondary ion mass spectra of glucose (M) obtained (a) from a liquid and (b) from a frozen mixture of glucose, NH_4Cl and glycerol (ratio of the weights 1:0.1:8).

Acknowledgement: Financial support of this work by the Deutsche Forschungsgemeinschaft is gratefully acknowledged.

References

1. M. Barber, R.S. Bordoli, G.J. Elliott, R.D. Sedgwick and A.N. Tyler, Anal. Chem. 54 (1982) 645 A

2. S.S. Wong and F.W. Röllgen, Nucl. Instr. Meth., in press

3. S.S. Wong and F.W. Röllgen, I. Manz and M. Przybylski, Biomed. Mass Spectrom. 12 (1985) 43

4. D.F. Barofsky, A.M. Ilias, E. Barofsky and J.H. Murphy, 32 nd Ann. Conf. on Mass Spectrom. Allied Topics, San Antonio 1984

5. D.F. Barofsky and E. Barofsky, Adv. in Mass Spectrom. 10, in press; these proceedings

6. W.V. Ligon and S.B. Dorn, Int. J. Mass Spectrom. Ion Proc. 57 (1984) 75

7. K.L. Busch, S.E. Unger, A. Vincze, R.G. Cooks and T. Keough, J. Am. Chem. Soc. 104 (1982) 1507

8. D.A. Kidwell, M.M. Ross and R.J. Colton, J. Am. Chem. Soc. 106 (1984) 2219

9. E. DePauw, Mass Spectrom. Rev., in press

10. L.K. Liu, K.L. Busch and R.G. Cooks, Anal. Chem. 53 (1981) 109

Secondary Ion Emission from Glycerol and Silver Supported Organic Molecules

M. Junack, W. Sichtermann, and A. Benninghoven
Physikalisches Institut der Universität Münster, Domagkstr. 75,
D-4400 Münster, F.R.G.

1. Introduction

Organic compounds with a low thermal stability and a low volatility have become accessible to mass spectrometry by analysing the secondary ions (SI) which are emitted during bombardment with ions or neutrals from the solid substance deposited on a substrate (e.g. Ag) [1-3] or from solutions in a low volatile liquid (e.g. glycerol) [4]. As characteristic ions $(M+H)^+$ and $(M-H)^-$ (M: molecule) and in case of salts the anions and cations are observed. Furthermore the SI-spectra exhibit ions due to attachment of atoms or molecules originating from the substrate (e.g. $(M+Na)^+$, $(M+Ag)^+$ or $(M+H+glycerol)^+$) and fragmentary ions.

In the following study, SI-spectra obtained from Ag and glycerol under comparable conditions of bombardment and coverage are presented, and then the enrichment, sputtering and regeneration occurring at the glycerol surface are investigated in more detail [5].

2. Experimental

As primary particles Ar^+-ions (energy 3 keV) were used, and the secondary ions were analysed by a quadrupol mass filter. For ionization from silver 1 µl of a solution of the compound was deposited on an Ag foil which had been etched in nitric acid. The solutions in glycerol were prepared without any additives. For the investigations in vacuum an amount of 10 µl was deposited in the cavity of a stainless steel sample holder (area 0.2 cm^2, depth 0.5 mm); the rate of evaporation was about 5 $\mu g cm^{-2} s^{-1}$ so that only a small part of the sample evaporated within the duration of the experiments of less than 1500 s.

3. Comparison of Secondary Ion Spectra

The positive SI-spectra of the surfactant Cetrimonium Bromide (CTAB) obtained from Ag and glycerol under static ion bombardment are shown in Fig.1a. In both cases the surface coverage is equivalent to 1 monolayer (for glycerol cf. Fig.2). The cation CTA^+ is emitted as molecular ion according to the salt-like structure of CTAB, and furthermore, ions originating from the substrate and typical fragmentations are observed due to losses of saturated hydrocarbon molecules from the side chain. The relative intensities of the characteristic ions are nearly the same for both preparations, but the absolute intensity from glycerol is about 10 times less than from Ag.

Similar results are obtained for the alkaloid atropine (Fig.1b). Again the coverage both on Ag and glycerol is equivalent to about 1 monolayer (for glycerol cf. Fig.3). The same specific ions ($(M+H)^+$ and fragments indicated by the mass numbers at the peaks) are observed in both cases, and also their relative intensities are comparable if the Ag substrate had been passivated by dipping in HI. On more reactive Ag (only etched or additionally cleaned by sputtering), the $(M+H)^+$ intensity is up to 100 times lower than shown in Fig.1b. The absolute

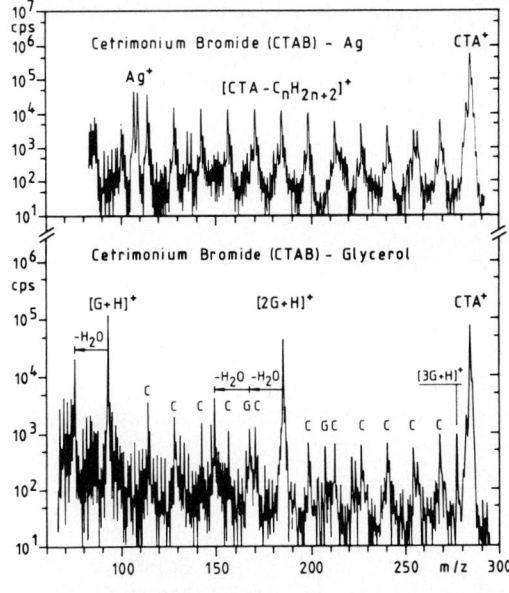

Fig.1 Positive SI-spectra of: (a) CTAB on Ag (0.1 nmol/cm^2) and in glycerol (G) (10^{-5} mol/l); (b) Atropine on passivated Ag (0.1 nmol/cm^2) and in glycerol (3×10^{-2} mol/l). Static primary ion bombardment, intensities normalized to 1 nA

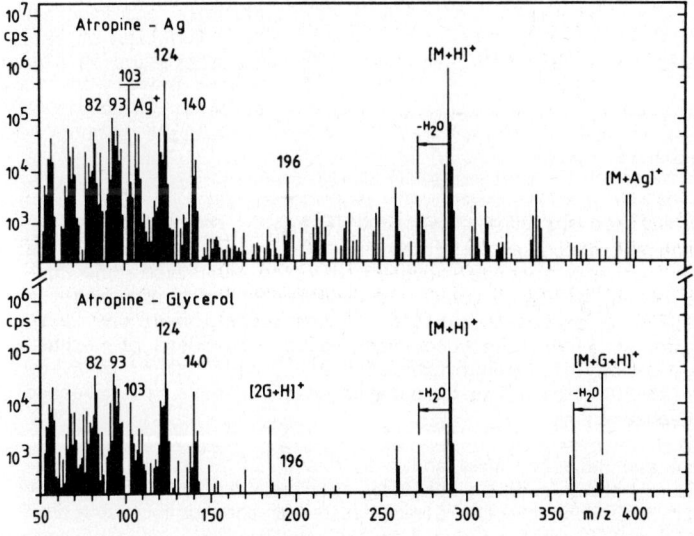

intensities obtained from glycerol are again nearly 10 times less than for the Ag substrate. In case of the peptide bradykinin, however, the spectra of samples with monolayer-coverages recorded with a time-of-flight mass spectrometer showed a little higher (M+H)$^+$ peak if a glycerol matrix was used instead of Ag. The negative SI-spectra contained in all cases investigated only negligible intensities of characteristic ions.

4. Secondary Ion Emission from Glycerol

4.1 Static Primary Bombardment

The dependance of the molecular ion intensities of different compounds on the concentration in glycerol is shown in Fig.2 and 3 for static primary bombardment.

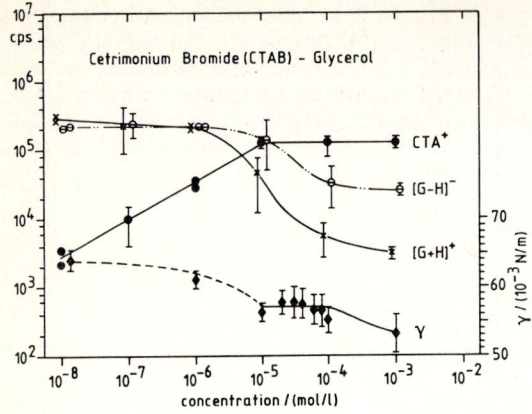

Fig.2 SI-intensities and surface tension γ versus the concentration of CTAB in glycerol (G). Static primary ion bombardment, intensities normalized to 1 nA

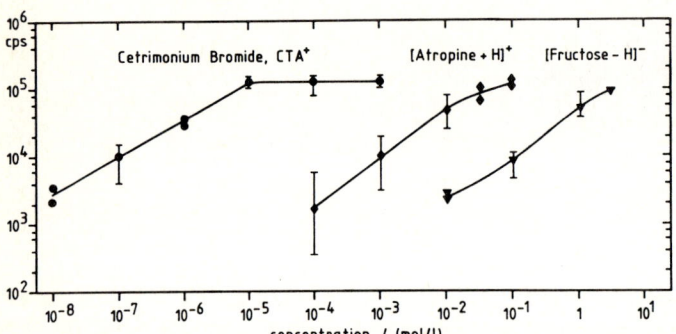

Fig.3 Molecular ion intensities of CTAB, atropine and fructose as a function of their concentration in glycerol. Static primary ion bombardment, intensities normalized to 1 nA

In case of CTAB a strong surface enrichment is found (Fig.2): The intensity of CTA^+ increases with the concentration until already at 10^{-5} mol/l a saturation is reached and the surface is nearly entirely covered, as it is seen from the decrease of the intensities of the glycerol ions. As expected for a surfactant, the surface tension of the solution decreases parallel to the formation of the surface layer and then remains at a relatively constant value. By means of Gibbs' equation of adsorption, a particle concentration of CTA^+-ions of 5×10^{13} cm^{-2} of the closed surface layer (at 10^{-5} mol/l) is calculated, which is also the coverage when the saturation is reached on Ag.

Atropine and fructose are detected only above 10^{-4} and 10^{-2} mol/l, respectively, in the same order as their surface activity decreases and the solubility in glycerol increases (Fig.3). For atropine a saturation of the surface coverage is reached at about 3×10^{-2} mol/l which is the limit of solubility, and as seen from the decreasing intensities of glycerol ions, the surface layer is nearly closed (see Fig.1b). In case of fructose, which has a structure similar to glycerol, high intensities of the $(M-H)^-$ ions and high coverages are obtained only at concentrations above 1 mol/l corresponding to the high solubility in glycerol.

4.2 Sputtering and Regeneration of the Surface

As shown in Fig.4 and 5 for CTAB the initial coverage may decrease at non-static primary bombardment. The exponential decrease of the CTA^+-intensity at 10^{-6} mol/l (Fig.4) confirms that as known for a surfactant the enriched surface layer has a monomolecular composition. The corresponding cross-section for desorption and damage of 3.2×10^{-14} cm^2 is quite the same as for sputtering from Ag (3.3×10^{-14} cm^2). After an initial decrease the CTA^+ intensity of the 10^{-5} molar

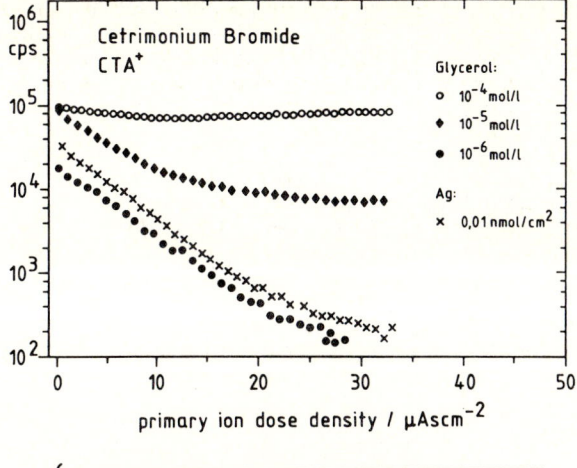

Fig.4 CTAB in glycerol: Intensity of CTA^+ versus the primary ion dose density. Primary current density 0.1 µAcm^{-2}, intensities normalized to 1 nA

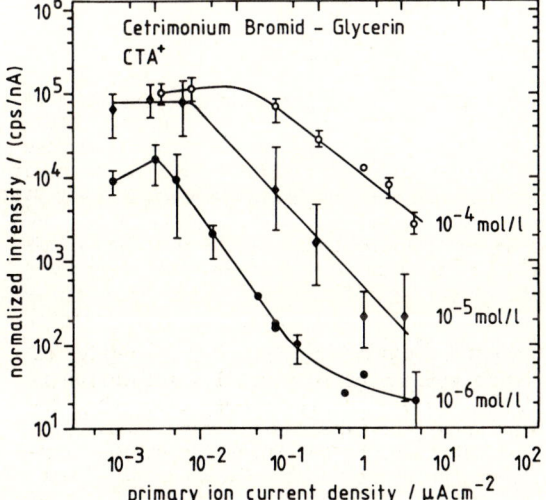

Fig.5 CTAB in Glycerol: Intensity of $\overline{CTA^+}$ in the quasi-equilibrium during bombardment versus the primary ion current density, intensities normalized to 1 nA

sample tends to a relatively constant value, while at 10^{-4} mol/l the coverage in this quasi-equilibrium state is nearly as high as in the beginning. At higher current densities, however, also for this concentration the coverage is lower during bombardment than under static conditions. As shown in Fig.5 the normalized intensity and thus the coverage in the quasi equilibrium strongly decreases at a given concentration with increasing primary ion current density. Inversely, it reincreases after lowering the bombardment. This shows that the regeneration of the surface is limited, and consequently the dependance of the intensity on the concentration is different for different primary ion bombardment.

The regeneration of the surface measured under static conditions after high prebombardment (15 µAcm^{-2} for 20 s) is shown in Fig.6 and 7 for solutions of CTAB and atropine. At 10^{-6} and 10^{-5} mol/l the surface of samples of CTAB is not entirely regenerated even after a delay of 1000 s in vacuum (Fig.6). The reincrease of the CTA^+ intensity is nearly proportional to $\sqrt{t}$ showing that in these cases the regeneration occurs mainly by diffusion and that the influence of the evaporation of glycerol is negligible, at least in the beginning. The coefficient of diffusion calculated from the increase of about 8×10^{-8} cm^2/s is in the same order of magnitude as for other compounds; it is at least 100 times less than for diffusion in water.

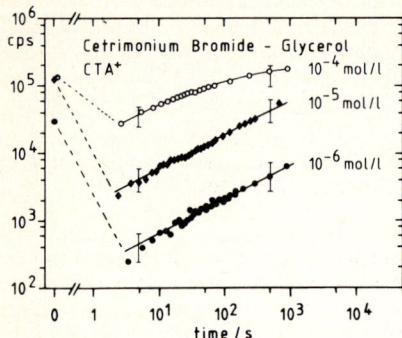
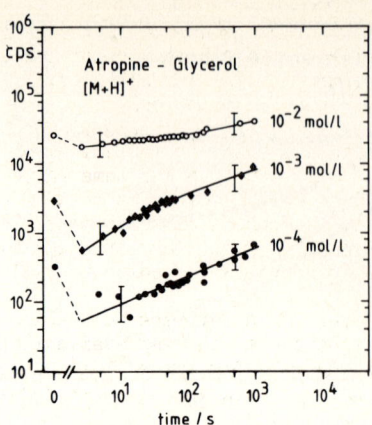

Fig.6 CTAB in glycerol: Intensity of CTA^+ as a function of the time of regeneration after prebombardment. The values at 0 s are the mean static equilibrium intensities taken from Fig.2. Static primary ion bombardment, intensities normalized to 1 nA

Fig.7 Atropine in glycerol: Intensity of $(atropine+H)^+$ versus the time of regeneration after prebombardment. The data at 0 s are the initial intensities before the bombardment. Static primary bombardment, intensities normalized to 1 nA

As seen already from Fig.4/5 the surface layer of 10^{-4} molar solutions is less affected by the primary bombardment than at lower concentrations, and furthermore it is rebuilt after about 200 s probably due to a higher rate of regeneration by evaporation of glycerol. For the same reason the influence of the prebombardment is much less for atropine than for CTAB at comparable surface coverages (Fig.7). Because of the higher concentrations, the surface of the 10^{-4} and 10^{-3} molar solutions is regenerated after about 200 s, and then the intensity even exceeds the initial value (given at 0 s) which can be explained by a concentration at the surface due to the evaporation of glycerol. From a comparison with the rate of evaporation it follows that there is also a back-diffusion into the bulk. Finally, samples with 10^{-2} mol/l show only a small influence of the prebombardment.

4.3 Sputtering Yields

Sputtering yields could be determined by combining the coverage measured during enhanced bombardment with the primary particle flux and a calculated rate of regeneration. In case of the CTA^+ ions a sputtering yield $S \approx 10-20$ is obtained assuming a regeneration by diffusion. For atropine the evaporation of glycerol can be considered as main process of transport to the surface because of the higher concentrations, and with this assumption also a value of $S \approx 20$ is calculated from the $(M+H)^+$ intensity. In contrast to the dissolved molecules about 140 molecules of glycerol are desorbed per primary ion, as determined from the increase of the desorption rate of glycerol during the bombardment. This higher value might be due to the lower binding energy or to thermal effects of the bombardment.

5. Conclusions

From the presented results it is seen that in vacuum the coverage at the surface of a solution e.g. in glycerol and therefore also the amount of substance accessible to ionization results from a superposition of several phenomena:

- Surface enrichment due to surface activity
- Sputtering by the primary bombardment
- Regeneration of the surface by diffusion and evaporation of the solvent
- Backdiffusion into the bulk

The single process of ion emission from glycerol is quite similar as for an Ag substrate as seen from the same value of the cross-section of CTA^+ and from the intensities, which differ at most by a factor of 10. Therefore also the total probability of transformation into a molecular ion should be comparable for both preparations. In fact, for $(atropine+H)^+$ a value of 3% was calculated for desorption from glycerol, which compares well to results obtained from Ag.

Thus the main difference between the emission from the solid state and a solution consists in the availability of the substance at the surface. The distribution over the whole volume of the liquid may lead, especially in case of well-soluble compounds, to a lower sensitivity than for a solid substrate.

References

1. A.Eicke, W.Sichtermann, A.Benninghoven, Org.Mass Spectrom. 15(1980)289
2. M.Junack, A.Eicke, W.Sichtermann, A.Benninghoven, in: A.Benninghoven (ed.), Ion Formation from Organic Solids, Springer Series in Chemical Physics, Vol.25, Springer-Verlag, Berlin 1982, p.177
3. R.J.Day, S.E.Unger, R.G.Cooks, Anal.Chem. 52(1980)557A
4. M.Barber, R.S.Bordoli, G.J.Elliot, R.D.Sedgwick, A.N.Tyler, Anal.Chem. 54(1982)645A
5. M.Junack, Ph.D.-Dissertation, Münster 1985

Temperature Effects in Particle Bombardment Mass Spectrometry of Methanol

R.N. Katz, T. Chaudhary, and F.H. Field
The Rockefeller University, 1230 York Avenue, New York, NY 10021, USA

The spectra produced by methanol when bombarded with a mixed beam of argon ions and atoms with a nominal energy of 8 keV have been obtained at temperatures between -94°C and -173°C. The freezing point of methanol is -98°C, so measurements were made on both liquid and solid methanol. The spectrum of the liquid at -94°C consists primarily of clusters of methanol around the proton, $H(CH_3OH)_n^+$, where the maximum value of n observed is 18. The spectrum of methanol at t= -104°C (below the freezing point) is very similar to that at t= -94°C. Thus freezing has no apparent effect on the spectrum. The spectrum of the same sample when cooled to -173°C is quite different, for the $H(CH_3OH)_n^+$ ions lose their prominance, and the high mass spectrum (above m/z 33) is without character, consisting of ions at more or less every mass. This change in the spectrum is reversible, for when the sample is warmed back to -107°C, the spectrum reverts to its original form with prominant $H(CH_3OH)_n^+$ ions. At an intermediate temperature (-136°C) the spectrum is a mixture of the features observed at -104°C and -173°C. Another way of representing this change in the character of the spectrum with temperature is to plot $\sum H(CH_3OH)_n$/TIC (TIC = total ion current) as a function of t, and such a plot shows a sharp decrease between t= -120°C -140°C. We have no explanation for this phenomenon. To investigate the possibility that the phenomenon could be related to surface melting of the solid produced by the ion bombardment, we measured the spectrum as a function of temperature at a bombardment intensity that was 7 times lower than that originally used. The same decrease in the relative concentration of $H(CH_3OH)_n^+$ clusters as the temperature decreased was observed.

A few incidental observations of the ionization behavior of impurity sodium and silver from the probe tip were made. No metal ions were ever observed when the probe tip was completely covered with either liquid or solid methanol. However, copious amounts of Na^+ and Ag^+ ions and their complexes with methanol were observed whenever the probe tip was partially covered with either liquid or solid methanol.

Internal Energy Distribution of Ions Emitted in Secondary Ion Mass Spectrometry

E. de Pauw, G. Pelzer, J. Marien, and P. Natalis

Liège University, Sart Tilman B6, B-4000 Liège, Belgium

I. INTRODUCTION

In secondary ion mass spectrometry, the extent of fragmentation is reduced when taking the spectra from solutions instead of using solid supports (1,2). Consequently, the internal energy of the ejected ions could either be different according to the physical state of the sample or be dissipated in another manner. In general, spectra obtained from solutions present peaks corresponding to both intact and fragmented species. An estimation of the internal energy of the sputtered ions could be made by the examination of their fragmentation in order to enlight the ion formation mechanism and the so-called matrix effect. In previous works (3,4,5), the extent of fragmentation for a given family of molecules has been related to substitution. The correlations found were explained in terms of the quasi-equilibrium theory and thermochemical considerations or in terms of mechanistic organic chemistry, using free energy relationships. The differences in fragment abundances were related to differences in the decomposition kinetics.

In electron impact studies, the situation is complicated by the fact that ions are produced from neutral molecules. During the ionization step of the experiment, undesired phenomena due to the substitution may arise that often obscure the results.

We present here the effect of the substitution on the liquid-SIMS spectra of benzyl-pyridinium and quaternary ammonium salts. Organic salts were chosen because they need no ionization, but only sputtering from the glycerol solution. In addition, in the case of benzyl-pyridinium cations, only one fragmentation occurs, at the charge-carrying site. A set of organic salts was elaborated by changing the nature and the position of the substituant on both pyridine and benzyl rings. Replacement of the pyridine moiety by a more basic radical was achieved by the synthesis of benzyl-triethylammonium salts. This set of cations is shown to be a good probe to study the internal energy of the cations after desolvation.

II. EXPERIMENTAL

The mass spectra were taken with an Extranuclear quadrupole mass spectrometer (7-162-8, mass range 1-1000 amu) previously described (6). The samples were prepared by mixing 1 mgr of the salt in 1 ml of glycerol. A Cs^+ primary beam (1 microampere measured on the target, 1 to 6 keV) was used to bombard the target.

The organic salts were prepared by condensation of the substituted benzyl halide on pyridine or triethylamine. In the case of substituents on the pyridine ring, dried ether was used as solvent. The salts precipitated after a few hours of reflux heating and were recrystallized in ether. Their purity was checked by the fusion point test.

The following compounds were synthetized:

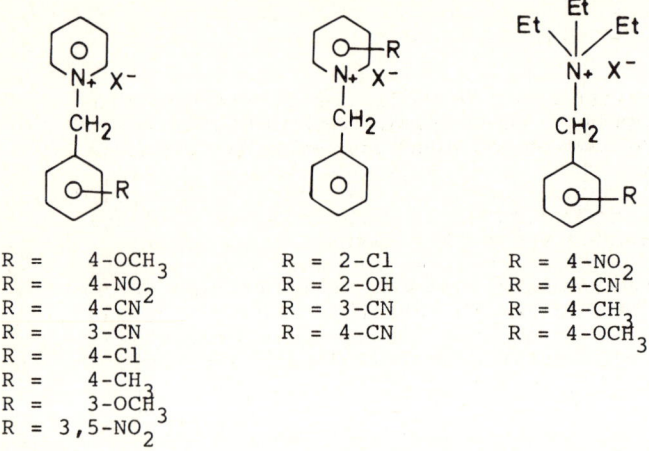

R =	4-OCH_3	R = 2-Cl	R = 4-NO_2
R =	4-NO_2	R = 2-OH	R = 4-CN
R =	4-CN	R = 3-CN	R = 4-CH_3
R =	3-CN	R = 4-CN	R = 4-OCH_3
R =	4-Cl		
R =	4-CH_3		
R =	3-OCH_3		
R =	3,5-NO_2		

III. RESULTS

A. Mass spectra of benzyl substituted benzyl-pyridinium salts

The mass spectra of benzyl-pyridinium salts present only two characteristic peaks in the mass region below the cation mass region, the dimer region will not be analysed. They correspond to the intact cation and the R-$C_7H_6^+$ ion. Their relative intensities are strongly influenced by the nature and the position of the substituent. A low intensity peak is also observed at mass 80, attributable to the protonated pyridine ion. The spectrum of the 4-methyl substituted salt is shown in figure 1. In table 1 is summarized the extent of

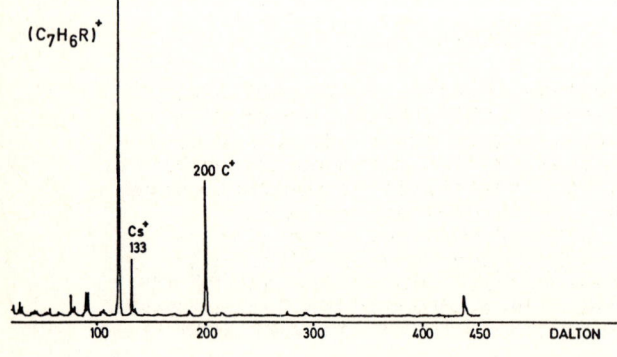

Figure 1. SIMS spectrum of 4-methoxybenzyl-pyridinium chloride (R= 4-OCH_3)

Table 1. F values for 4-Substituted benzyl-pyridinium salts

Substituent	F value	D IP	SIG+
H	0.50	0	0
4-Cl	0.43	0	0.12
4-CN	0.19	0.4	0.66
4-NO_2	0.05	0.6	0.79
4-CH_3	0.44	-0.3	-0.31
4-OCH_3	0.70	-0.7	-0.78
3-CN	0.24	----	0.56
3-OCH_3	0.44	----	-0.27

fragmentation represented by the relative intensities F of the R-$C_7H_6^+$ ions against the total due to the target compound. The values are calculated according to the formula

$$F = \frac{I(T_p)}{I(T_p) + I(C)}$$

where $I(T_p)$ and $I(C)$ are respectively the intensity of the fragment and the intact cation peaks.

B. Mass spectra of pyridine-substituted pyridinium salts

Mass spectra of benzyl-pyridinium salts substituted on the pyridine show the occurrence of the same fragmentation at the C-N bond. The fragmentation extent values for the other related compounds are summarized in table 2.

Table 2. F values for pyridine-substituted benzyl pyridinium salts

Substituent	F value	D IP
H	0.5	0
4-CN	0.65	0.64
2-Cl	0.48	0.14
3-CN	0.62	0.54
2-OH	0.27	-0.55

C. Mass spectra of benzyl-triethylammonium salts

The mass spectra of benzyl-triethylammonium salts are composed of the intact cation and the substituted benzyl peaks. The C-N bond breaking is again observed. In addition, as in the case of pyridinium salts, the nitrogen-containing radical is present, but now its characteristic peaks are more intense. One observes peaks at 100, 101 and 102 amu, respectively, attributable to the protonated triethylamine ion $(M+H)^+$, the corresponding radical ion $M^{\cdot+}$ and the iminium ion $(M-H)^+$. Their relative intensity also depends on the substitution. The fragmentation extent values are summarized in table 3.

Table 3. F values for benzyl-triethylammonium salts

Substituent	F value	D IP
H	0.33	0
4-CN	0.07	0.4
4-NO$_2$	0.06	0.6
4-CH$_3$	0.40	-0.3
4-OCH$_3$	0.61	-0.7

IV. DISCUSSION

The fragmentation extent has been correlated for each family of compounds to the sigma+ constant, describing the ability of the transition state to stabilize a positive charge. It has also been correlated to a more direct energetic parameter, i.e. the ionization energy change (D IP) for the corresponding toluene or pyridine ring upon substitution. In both cases, a very good correlation ($r^2 \geq 0.9$) has been obtained. This is illustrated in figure 2 where the correlation with sigma+ and D IP are shown for the substituted benzyl-pyridinium salts. On the other hand, no correlation was obtained between the methylene carbon C^{13} NMR chemical shift and the sigma+ or D IP values.

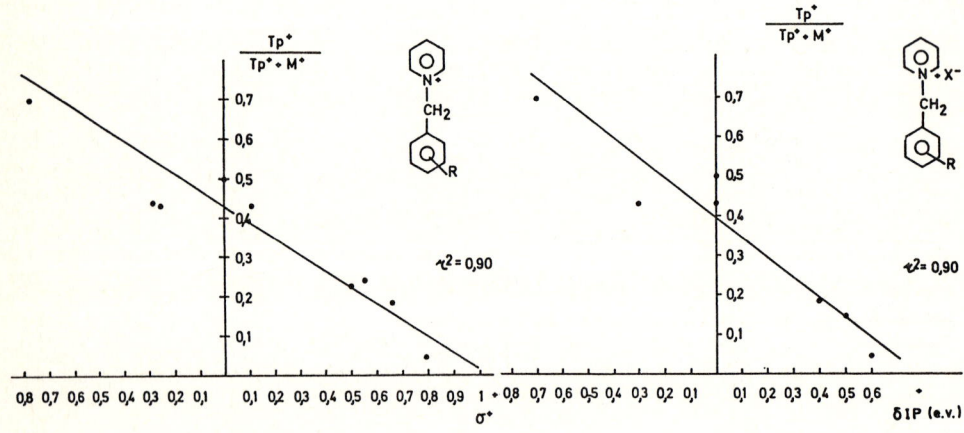

Figure 2. Correlations between the F values and substituent parameters

The existence of a correlation similar to the correlation observed for the thermal reactivity of the cations in solution allows us to make some basic assumptions concerning the ion formation processes:

1) The energy deposition is independent of the nature of the solute.
2) The dissociation is induced, in the case of preformed ions, by vibrational excitation of the ion ground state. In contrast to what happens in electron impact, no selection

rule is involved in the process of energy uptake by the solute, and linear free energy relationships are expected to be more valid.

3) The impact energy must be dissipated in such a way that ions with low internal energy content finally undergo dissociation, so that small changes in the fragments stability have large effects on the F values. This is performed by energy buffering via the matrix, and could consist in successive desolvation steps of large clusters.

We will now try to use the fragmentation extent correlation with substituent properties to evaluate the internal energy distribution of the fragmenting species. If only one fragmentation is observed, the cation can be approximated to a diatomic closed shell ion $(Pyr-Tp)^+$ in which one atom is changed by substitution on a ring. In the total energy diagram of a diatomic species, this results taking the vibrational ground state as reference, to a change in the dissociation limit position, in our case, for the C-N vibration mode. For an energy deposition only matrix dependent, the energy uptake will be the same for solutes of the same family. The total energy curves for the various cations can then be plotted in front of the unknown energy uptake curve, and the F values will be used to reconstruct its shape. In fact, each F value will correspond to the area of the energy uptake curve above the corresponding dissociation limit, normalized by the total area under this curve. This is represented in figure 3. A first approximation of the internal energy distribution function has been made on a gaussian basis. The agreement with the F values fits within 15% and gives a half-width value of 1.3 eV

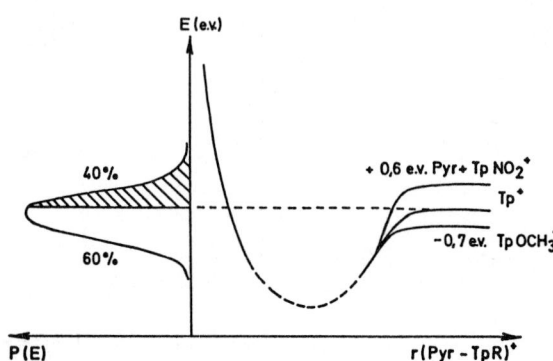

Figure 3. Tentative internal energy distribution curve of fragmenting cations

The present model is of course a first approximation, and it gives only the energy distribution on the vibrational mode-inducing dissociation. A study of the metastable ions abundances and of the variable energy collision-induced dissociations is under progress.

Finally, such a set of organic salts, whose fragmentation extent is, in a predictible way, sensitive to energy deposition, could be employed as a probe to compare the internal energy content in the various so-called soft ionization techniques.

Bibliography

1. A.Ba-Isa, K.L.Bush, R.G.Cooks, A.Vincze and I.Granoth
   Tetrahedron, 39, 1983, 591
2. S.S. Wong and F.W.Roellgen
   Nucl.Inst.Meth., in press
3. M.M.Bursey and F.W.Mc Lafferty
   J.Am.Chem.Soc., 88, 1966, 529
4. R.S.Ward, R.G.Cooks and D.H.Williams
   J.Am.Chem.Soc., 91, 1969, 2727
5. F.W.Mc Lafferty, T.Wachs, C.Lifshitz, G.Innorta and P.Irving
   J.Am.Chem.Soc., 92, 1970, 6867
6. G.Pelzer, E.De Pauw, D.V.Dung and J.Marien
   J.Phys.Chem., 88, 1984, 5065

Fast Atom Bombardment of Peptides Above 5000 Daltons

C. Fenselau and K. Hyver

Department of Pharmacology, Johns Hopkins University School of Medicine, 725 North Wolfe, Batimore, MD, USA

This paper summarizes analyses and investigations made through the last four years in a sector mass spectrometer with a 23 Kg magnet, transmitting ions up to about 3000 Daltons at full accelerating voltage. In order to work with ions above 5000 Daltons the accelerating voltage and often the resolution were lowered. Many of the considerations and much of the protocol we have developed to push this instrument to its limit will be applicable to higher mass magnets as these become available. The Kratos MS-50 used in this work was fitted with a saddle field gun for fast atom bombardment, an 8kV post accelerating detector, and the Kratos DS-55 data system with software written both in house and commercially. This instrument was routinely calibrated to assign masses in real time to about 4500 Daltons. Alternatively, the computer system could be calibrated through 3000 amu anywhere in the mass range up to at least 7500 Daltons for off-line mass assignment. Thus masses above 5000 were assigned either by peak matching or by computer supported off-line comparison to cesium iodide reference spectra.

The increasing complexity of molecular ion multiplets encountered above 5000 Daltons has been discussed [1]. The molecular ion group of human proinsulin, is symmetrical and has nearly the same peak width at half-height as the molecular ion group in the same mass range of polystyrene oligomer n = 90, even though one formula contains one heteroatom for every two carbon atoms and the other contains no heteroatoms. Molecular ion multiplets in this mass range no longer have very distinctive shapes. These and other considerations [2] have led us to propose [1] that accurate average masses are the molecular weights most reliably reported for compounds in the mass range above 5000 Daltons. Average masses can be confounded by mixtures of M^+, $(M+H)^+$, $(M-H)^+$, etc. Selection of a suitable matrix appears to permit $(M+H)^+$ species to be maximized.

Figure 1 shows positive ion spectrum of an unknown preprosomatostatin peptide. The wide mass or survey scan was measured by time binning [3], in which the computer sums all ions detected through designated time intervals while the magnetic field is scanned. About 1000 such bins compose the spectrum in Fig. 1. This approach provides the sensitivity and speed required to acquire such a wide scan, and also improves the signal-to-noise ratio. Average mass assignments were made off-line by comparison to reference spectra. The baseline is elevated, especially in the lower mass range, reflecting extensive incoherent fragmentation. The baseline is drawn as an envelope, however the signal actually drops to zero between each mass. Protonated molecular ions (and, in other cases, polyprotonated polyvalent molecular ions) appear distinctively above the baseline. Sequence ions are not discernible, in contrast to the situation in FAB spectra of peptides of lower molecular weights [4]. This same lack of discernible sequence fragmentation is also a feature of spectra of heavier peptides ionized by megavolt particles [5-8]. At last one theoretical discussion predicts this change in the nature of fragmentation of heavier compounds [9], and predicts that small peripheral groups will be lost preferentially. Losses of this kind have been confirmed experimentally [10].

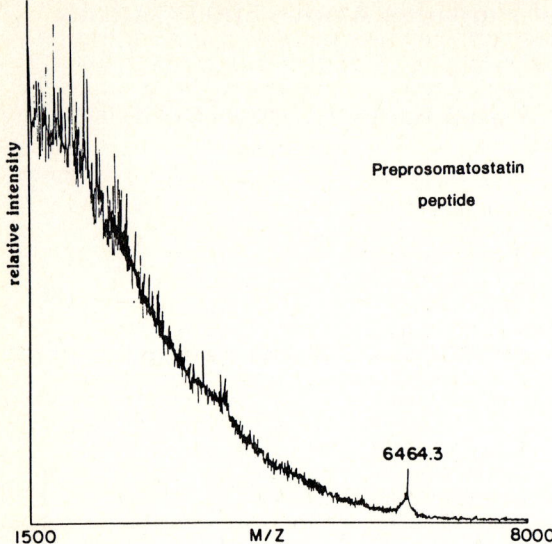

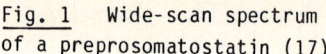

Fig. 1 Wide-scan spectrum of a preprosomatostatin (17)

Our protocol for analysis of middle molecules involves first obtaining a wide mass or survey scan. Subsequently, masses ranges recognized to be of interest can be examined with higher resolution [3,11].

Figure 2 presents one of the molecular ion multiplets of mouse epidermal growth factor, along with cesium perfluorohexanesulfonate cluster reference ions [12], both recorded at 6000 resolution. The cluster ion peaks show characteristically remarkable intensities. These cluster ions contain no hydrogen, and consequently $(M-H)^+$ ions need not be considered. Some $(M-19)^+$. ions are detected at the far right [2].

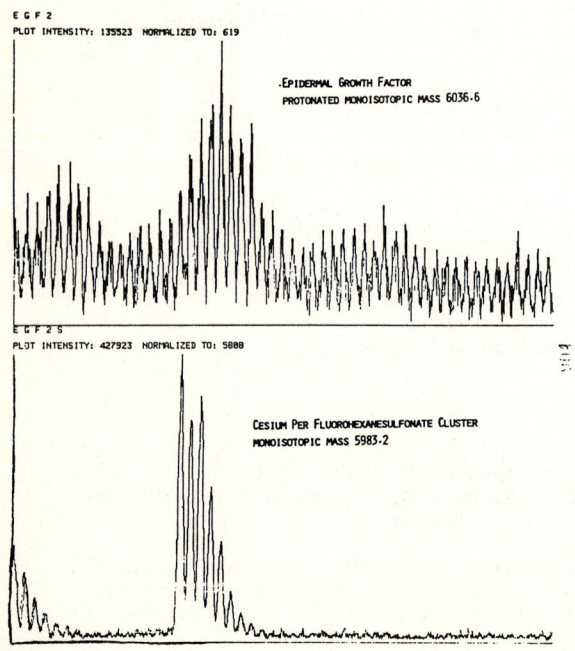

Fig. 2 Partial spectra of epidermal growth factor and cesium perfluorohexane sulfonate at 6000 resolution.

The mass spectral characterization of this factor as a mixture of two peptides differing by one asparagine moiety has been reported in the biochemical literature [11]. When the factor was characterized ten years ago, [13] each of the amino acid sequence steps produced a mixture. However, uncertainty about the completeness of the Edman reaction confounded interpretation of these mixtures.

The challenge to generate fragment ions containing sequence and other structural information from these heavier peptides is currently being addressed. However, the molecular weight information which is available is very useful. Examples of the kinds of peptide problems which can be addressed include surveying peptide molecular weight differences as a function of species, confirmation of chemical alterations of peptides, molecular weight determinations on unknown peptides including tryptic mixtures, as well as qualitative [12,14] and quantitative (15) evaluations of heterogeneity.

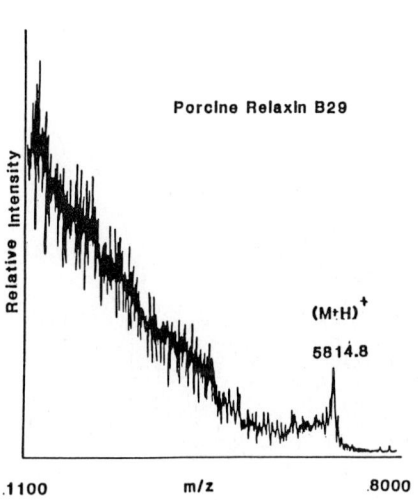 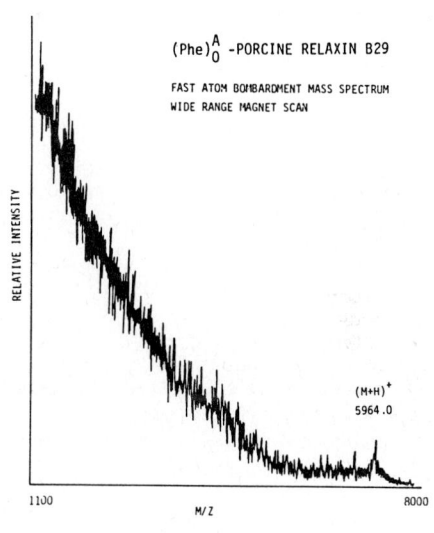

Fig. 3 Wide-scan spectrum of B29 porcine relaxin.

Fig. 4 Wide-scan spectrum of Phe A0 B29 porcine relaxin (16)

Figures 3 and 4 present wide range mass spectra of porcine relaxin containing twenty-nine amino acids in the B chain, and a chemically altered variant predicted to carry an additional amino acid in the A chain [16]. The successful addition of phenylalanine is confirmed by the increase in the average molecular weight. The masses of the A and B chains may be distinguished by reductive cleavage in the FAB matrix just prior to mass spectral analysis. A number of other chemically altered relaxin derivatives have also been characterized by FAB mass spectrometry, as indicated in the Table.

An example of the molecular weight determination of an unknown sample is shown in Fig. 1. The sequence of the preprosomatostatin is currently under study [17]. All hydrolysis and sequence experiments can now be interpreted in the context of this average mass determination.

The Table lists a number of average molecular weights determined from these time-binned spectra, and also the average molecular weights predicted. Overall the accuracy in the Table is good, and particularly so when compared to the accuracy provided to biochemists by gel electrophoresis.

TABLE 1 AVERAGE MOLECULAR WEIGHTS OF PEPTIDE HORMONES MEASURED FROM WIDE SCANS ON THE MS-50

	Theoretical (MH)$^+$	Experimental (MH)$^+$
Insulin, Bovine	5735.4	5736.0
Porcine (3)	5778.6	5777.7
Mouse Epidermal Growth Factor (11)		
EGF	6040.6	6040.3
EGF	5926.5	5926.8
Porcine Relaxin		
B27	5602.6	5602.7
B29	5815.8	5814.8
Phe A0 B29	5963.0	5964.0
des Arg A1 B29	5659.6	5661.3
des Arg Met B29	5528.4	5528.8

It seems likely that fast atom bombardment can be used to analyse heavier molecules than those discussed here. This will be confirmed by the use of FAB or liquid SIMS with the next generation of sector magnets, designed to transmit ions over 10000 Daltons, and with time-of-flight analyzers. High-flux liquid SIMS has been recently combined with time-of-flight mass spectrometry [18].

1. J. Yergey, D. Heller, G. Hansen, R.J. Cotter and C. Fenselau: Anal. Chem. 55, 353 (1983).
2. C. Fenselau, J. Yergey and D. Heller: Intern. J. Mass Spectrom. Ion Physics 53, 5 (1983).
3. R.J. Cotter, B.S. Larsen, D.N. Heller, J.E. Campana and C. Fenselau: Anal. Chem. 57, 1479 (1985).
4. K.L. Rinehart: Science 218, 254 (1982).
5. R.D. MacFarlane: Biomedical Mass Spectrometry, ed. G.R. Waller and O.C. Dermer (Wiley Interscience, New York, 1980).
6. B. Sundqvist, I. Kamensky, P. Hakansson, J. Kjellberg, M. Salehpour, S. Widdiyasekera, J. Fohlman, P.A. Peterson and P. Roepstorff: Biomed. Mass Spectrom. 11, 242 (1984).
7. B.T. Chait and F.H. Field: Intern. J. Mass Spectrom. Ion Processes 65, 169 (1985).
8. M. Alai, C. Fenselau, R.J. Cotter and C. Schwabe: Anal. Chem. in press.
9. D.L. Bunker and F.M. Wang: J. Amer. Chem. Soc. 99, 7457 (1977).
10. P. Demirev, M. Alai, R. van Breemen, R. Cotter and C. Fenselau: Proceedings of the Fifth International Conference on Secondary Ion Mass Spectrometry, Springer-Verlag Series in Chemical Physics, 1986.
11. K.J. Hyver, J.E. Campana, R.J. Cotter and C. Fenselau: Biochem. Biophys. Res. Commun. 130, 1287 (1985).
12. D.N. Heller, C. Fenselau, J. Yergey and R.J. Cotter: Anal. Chem. 56, 2274 (1984).
13. C.R. Savage, T. Inagami and S. Cohen: J. Biol. Chem. 247, 7612 (1972).
14. C. Fenselau, D.N. Heller, M.S. Miller and H.B. White: Anal. Biochem. 150, (1985).
15. R.R. Townsend, D.N. Heller, C.C. Fenselau and Y.C. Lee: Biochemistry 23, 6389 (1984).
16. Christian Schwabe, Department of Biochemistry, Medical University of South Carolina
17. Philip Andrews, Department of Biochemistry, Purdue University.
18. R.J. Cotter: Anal. Chem. 56, 2594 (1984).

Supported by grant PCM 82-09954 from the National Science Foundation.

Amino Acid Sequencing of Peptide Mixture: Structural Analysis of Human Hemoglobin Variants (Digit Printing Method)

T. Matsuo[1], T. Sakurai[1], I. Katakuse[1], H. Matsuda[1], Y. Wada[2], and A. Hayashi[2]

[1] Institute of Physics, College of General Education, Osaka University, Toyonaka 560, Japan
[2] Osaka Medical Center and Research Institute for Maternal and Child Health, Izumi 590-02, Japan

Abstract

Two strategies for determining the amino acid sequence of abnormal peptide containing amino acid substitution were developed. One is the mass spectrometric analysis of the peptide mixtures obtained by different kind of enzymatic digestions. Another is the combination of the sequenator with mass spectrometry. Two new abnormal fetal hemoglobin variants were determined by these techniques.

1. Introduction

Rapid advance in the mass spectrometry of heavy molecular compounds since 1970 depends on the following two developments: One is the new ionizing technique, field desorption (FD)[1], secondary ionization (SIMS)[2] and fast atom bombardment (FAB) [3] and another is the new powerful mass spectrometers with large magnetic rigidity [4]. We have applied these techniques to the sequencing of polypeptide. This project has proceeded along the following four steps.

In the first step, we analyzed various kinds of known single peptide by FD ion source using silicon emitter [5].

In the second step, we analyzed a complex mixture of peptides, and obtained clear FD mass spectra of the components. We immediately recognized that mass spectrometry can determine the molecular weights of the components without pre-separation.

In the third step, we proposed the new sequencing method combined with Edman degradation at 8th International Mass Spectrometry Conference in 1979 [6-8]. This method is now widely used, especially in a combination with SIMS or FAB ionization techniques.

In the fourth step, we established a method for detecting a polypeptide containing amino acid substitution. Since this mass spectrometric approach provides digitalized data on peptide analysis, we call it "digit printing method"[9]. We have applied this method to the structural analysis of abnormal globins and succeeded in determining the position and the type of substitution of two new abnormal fetal hemoglobin variants.

2. Experimental

2.1 Sample preparation

The proband was found in a mass screening study for fetal hemoglobin variants in Osaka. The screening method was isoelectric focusing of globins from the dried blood spot on filter paper. Hemoglobin was purified by DEAE column chromatography, globins were purified by CM cellulose column chromatography in urea, and aminoethylated. Tryptic digestion of the aminoethylated globin was carried out with 2% (w/w) of TPCK-trypsin (worthington) in 0.05 M ammonium bicarbonate (pH 7.8) at 37°C for 1h.

2.2 Mass spectrometer

Mass spectra were obtained using a Matsuda-type mass spectrometer (magnet radius 0.5 meter; acceleration voltage 6kV; resolving power 2000). Mass data were analysed by a DA 5000 data analysis system on line (JEOL Co. Ltd). A conventional FD ion source was modified to a SI mass spectrometric ion source by mounting a simple primary gun of cold cathode discharge type. The discharge voltage and current of the primary gun were 9 kV and 0.5 mA, respectively. The current hitting the sample tip was 5 μA.

Each sample was dissolved in water/methonol (1:1) at a concentration of 5μg/μl. About 2 μl of dissolved sample was loaded on a sample tip with 2 μl of glycerol and 0.5 μg of trichloroacetic acid (TCA) in methanol solution.

2.3 Sequenator

Amino acid sequences of separated peptide fragments were determined by a gas-phase sequenator (Applied Biosystems).

3. Results and discussion

3.1 Hb-F Izumi

Mass spectra of a tryptic digest of the normal γ globin and that of abnormal γ globin are shown in Fig. 1. By comparing the corresponding peaks carefully, it became clear that the absence of the peak T-1,2 at m/z 1919 and the appearance of

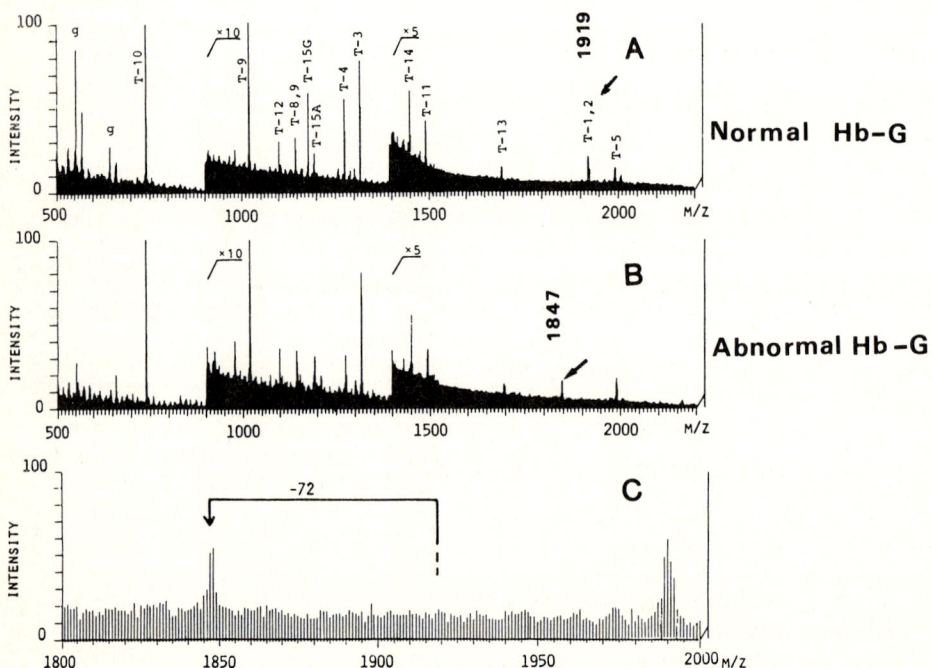

Fig. 1 Mass spectra of a tryptic digest of the normal γ globin (A) and the abnormal γ globin (B) spectrum C shows the mass range of m/z 1800 - 2000 of the spectrum B.

a new peak at m/z 1847. The decrease by 72 mass unit could be explained by only two types of substitution (Glu→Gly) or (Trp→Asp). Because the abnormal γ globin had a higher isoelectric point than that of normal γ globin, the substitution (Glu→Gly) is more probable. Peptide fragment T-1, 2 of normal γ globin contains two glutamic acid at 5th and 6th positions. For the purpose of determining the position, an additional cleavage with V8 protease was carried out on the tryptic digest. The mass spectrum of this new peptide mixture determined the substitution at the 6th position.(data not shown) From the mass spectrum shown in Fig. 1, we knew that the variant was originated in the A_γ globin gene because T-15G (m/z = 1178) was missing in the Fig. 1B. The primary structure of the abnormal hemoglobin was determined to be A_γ ^{6}Glu→Gly and it was named HbF Izumi.

3.2 Hb-F Yamaguchi

Mass spectrum of this abnormal globin is shown in Fig. 2. The abnormality is detected in peptide T-10. The molecular weight of T-10 of abnormal globin is one mass unit smaller than that of normal one. The substitution (Glu→Gln) or (Asp→Asn) is expected. As shown in Table 1, T-10 fragment of normal globin contains two aspartic acids (at 79th and 80th). In order to determine the position, Edman degradation of tryptic mixture of abnormal γ globin was carried out. The tryptic peptides of abnormal γ globin were fractionated by reserved phase HPLC as shown in Fig. 3. The complete separation of normal T-3 and abnormal T-10 is very difficult. These two components were collected in a same fraction. By taking the mass spectra of this fraction shown in Fig. 4, the molecular weights of component peptide can be determined to be 738 and 1315 respectively. Then, this fraction was analyzed by an automatic sequenator. From the result of sequenator given in Table 2, it was concluded that the substitution was at the 80th position. The primary structure was determined to be 80 Asp→Asn and it was named Hb F Yamaguchi.

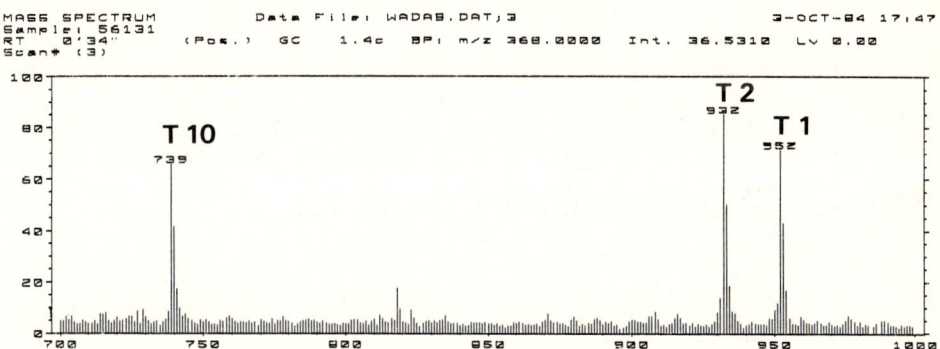

Fig. 2 SI Mass Spectrum of tryptic peptides of abnormal γ globin

Table 1 Amino acid sequence of normal T-10 (sequence I), T-10 with a Asp→Asn substitution at 79th position (sequence II), T-10 with a Asp→Asn substitution at 80th position (sequence III) and normal T-3 (sequence IV).

Position No. →	77	78	79	80	81	82			
I	His	Leu	Asp	Asp	Leu	Lys		M.W.	739
II	His	Leu	Asn	Asp	Leu	Lys		M.W.	738
III	His	Leu	Asp	Asn	Leu	Lys		M.W.	738

Position No. →	18	19	20	21	22	23	...		
IV	Val	Asn	Val	Glu	Asp	Gly	...	M.W.	1315

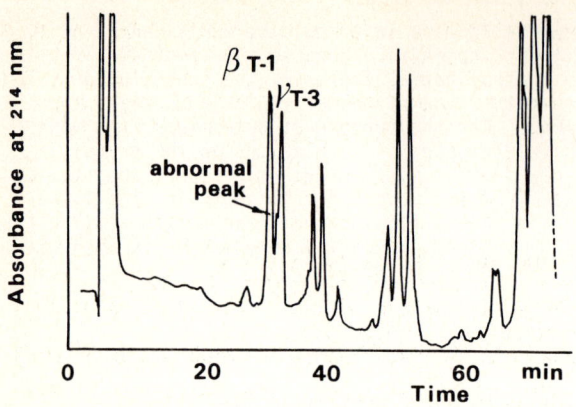

Fig. 3 HPLC chromatogram of tryptic digest of globins from a Hb F Yamguchi patient.

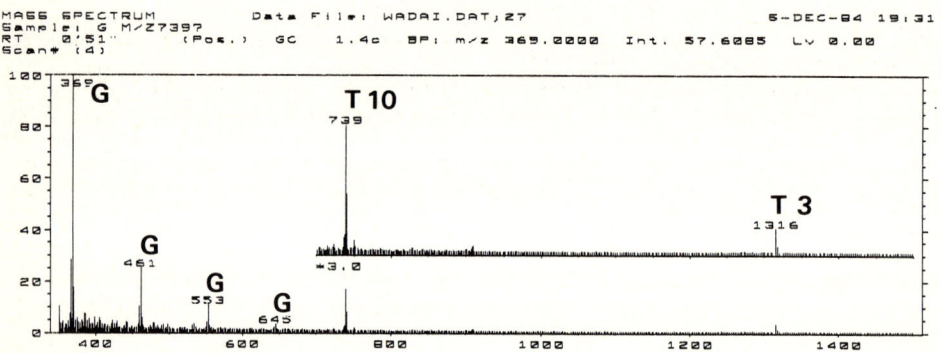

Fig. 4 Mass spectrum of the roughly fractionated peak which contain T-3 and T-10.

Table 2 List of released PTH amino acids from a mixture of abnormal γ T-10 and normal T-3.

No. of degradation	1	2	3	4	5
PTH amino acids	His, Val	Leu, Asn	Asp, Val	Asn, Glu	Leu, Asp

4. Conclusion

We have developed two methods for characterizing protein variants. The first one is to determine the position and the type of amino acid substitution from the mass spectra of peptide mixtures obtained by different kind of enzymatic digestions. (The case of Hb-F Izumi). Since this mass spectrometric approach provides digitalized data on a peptide mapping or a finger printing of peptides, we call it *"digit printing method"*.

The second one is to combine the mass spectrometric technique with Edman degradation. The distinguishing feature of this method is the possibility of sequencing of peptide mixture without isolation of a single peptide. Edman degradation can be performed for peptide mixture. From the list of released PTH amino acids shown in

Table 2, the position and the type of substitution can be determined (the case Hb-F Yamaguchi).

These two methods are useful for determining different kinds of amino acid mutations, depending on the sample.

References

1. H.D. Beckey: Int. J. Mass Spectrom. Ion Phys. 2, 500 (1969).
2. A. Benninghoven, D. Jaspers and W. Sichtermann: Appl. Phys. 11, 35 (1976).
3. M. Barber, R.S. Bordoli, R.D. Sedgwick and A.N. Tyler: Annual Conf. on Mass Spectrom. and Allied Topics. Minneapolis, 351 (1981).
4. H. Matsuda: At. Mass Fundam. Constants 5, 185 (1976) and commercial high field mass spectrometers.
5. T. Matsuo, H. Matsuda and I. Katakuse: Anal. Chem. 51 69 (1979).
6. T. Matsuo, H. Matsuda and I. Katakuse: Adv. in Mass Spectrom. 8, 990 (1980).
7. Y. Shimonishi, Y-M Hong, T. Matsuo, H. Matsuda and I. Katakuse: Chem. Lett. 1369 (1979).
8. T. Matsuo, I. Katakuse, H. Matsuda, Y. Shimonishi, Y-M Hong and Y. Izumi: Mass Spectroscopy (Japan) 28, 169 (1980).
9. Y. Wada, A. Hayashi, M. Fushimura, I. Katakuse, T. Ichihara, H. Nakabushi, T. Matsuo, T. Sakurai and H. Matsuda: Biochem. Biophys. Acta. 749, 244 (1983).

Oligonucleotide Sputtering from Liquid Matrices

L. Grotjahn

GBF (Gesellschaft für Biotechnologische Forschung mbH),
Mascheroder Weg 1, D-3300 Braunschweig, F. R. G.

1 Introduction

During the past few years the chemical synthesis of DNA fragments has become progressively more important for the modern biosciences. This has been made possible by major progress in synthesis technology, in particular the use of better coupling and deprotection methods, the employment of suitable solid supports, and refined isolation and characterization techniques.

Such molecules serve as "Linkers and Adaptors" in recombinant DNA techniques, and allow a vast number of new combinations of gene material to be produced. Other synthetic DNA fragments are used to perform site-directed mutagenesis of natural DNA or to achieve total synthesis.

For a long time mass spectrometry has been used to try to characterize such compounds, either as the oligomeric building blocks or the final free oligonucleotides.

The first investigations with field desorption on free oligonucleotides [1,2] only gave results up to the dimer.

Fully protected oligonucleotides were first succesfully investigated with time-of-flight instruments, using either 252-Californium plasma desorption mass spectrometry [3,4] or secondary ion mass spectrometry (SIMS) [5,6].

Investigations with static SIMS gave results for both the fully protected oligonucleotide up to four base units in length and free oligonucleotides up to 6 base units in length [7]. In the negative ion mode one can see both molecular ions and sequence ions. The method is limited by the preparation technique. The bonding energies between the oligonucleotides and the metal target increase with the number of nucleotide units and cause a decrease in the intensity of the secondary ions. However, when the preparation technique was changed and glycerol was used as the matrix, these problems do not occur. This, used in conjunction with a primary neutral atom beam on a high field mass spectrometer, the so-called technique of fast atom bombardment (FAB) mass spectrometry in the negative ion mode, is a suitable method for characterizing mononucleotides and for sequencing the oligomeric building blocks of the phosphotriester synthesis, as well as for sequencing the final end-product [8-14].

2 Experimental

All FAB mass spectra in the present article were obtained on a KRATOS MS 50 S mass spectrometer with a high-field magnet equipped with a KRATOS FAB source. The atom gun used xenon and produced a beam of neutral atoms at 8-9 kV. The spectra were recorded at a magnet scan rate of 300 s/dec or in the ESA scan mode (3 sec sweep with a peak switch of 10000 ppm) by accumulation with a multichannel analyzer TRACOR TN 1500.

The oligodeoxyribonucleotides were synthesized following, in general, the established phosphate triester procedure [9]. The ribonucleotide were of commercial origin (P-L Biochemicals, Milwaukee, USA).

Samples were prepared in various ways. A water solution of the triethylammonium salt of each free oligonucleotide, ca. 4 ul, containing ca. 1 OD_{260} (ca. 10 nmol), was injected into the glycerol matrix (ca. 2 ul). The negatively charged protected oligonucleotide (triethylammonium salt) were directly introduced into a 1:1 mixture of glycerol and triethylene glycol.

3 Results and Discussion

Protected oligonucleotides with a negative charge at the terminal phosphate, produced either by initial chemical degradation of the terminal phosphate protecting group or by reaction in the matrix (for example ethanolamines), generally yield the sequence information by consideration of the 3'-phosphate sequence ions. The shorthand structure and the relevant regions of the protected tetramer TGAC are shown in fig. 1. G is protected with an isobutyryl, A with a benzoyl, C with an anysoyl, the phosphate with an orthochlorophenyl and the 5'-oxygen with a monomethoxytrityl group. From the examination of this set of ions the base sequence of this oligomeric building block is easily deduced.

The fragmentation behaviour of the fully deprotected oligonucleotide allows a bidirectional sequence analysis [9]. Fig. 2 shows the

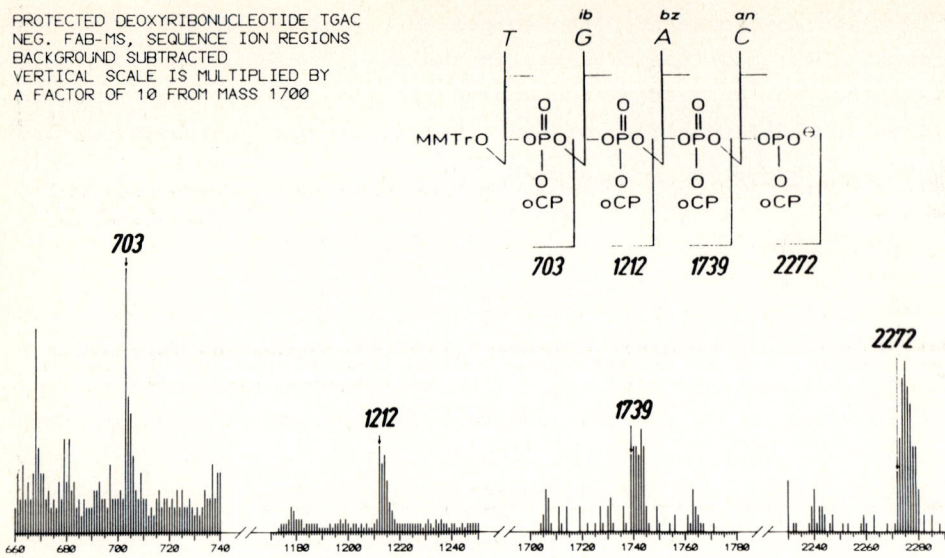

Fig. 1 Relevant regions of the negative ion FAB mass spectrum of negatively charged protected d(TGAC)

negative ion FAB mass spectrum of the deoxyribonucleotide GAAGCTTC. The deprotonated molecular ion is found at m/z 2407. The shorthand formula indicates the masses, which were formed by cleavage of the bond between the carbon of the sugar moiety and the phosphate oxygen.

The main fragmentations result in the formentation of sequence ions with either 3'- or 5'-terminal phosphates (3'- or 5'-phosphate sequence ions). A significant distinction was found for the two different types of phosphate sequence ions, 5'-phosphate sequence ions are always more abundant than the corresponding 3'-phosphate sequence ions with the same number of nucleotide units. This simple fragmentation behaviour allows an independent sequencing of the oligonucleotide chain either from the 3'- or from the 5'-end (bidirectional sequence analysis).

Often smaller m/z -18 ions (loss of water) and m/z +80 ions (perhaps +HPO_3 artefacts) arise next to the sequence and matrix ions; but these do not disturb the sequence analysis. The less intense 3'-phosphate sequence ions (from the 5'-end) permit determination of the sequence information in the correct direction GAAGCTTC whereas the more intense 5'-phosphate sequence ions (from the 3'-end) permit determination from the opposite end.

Figure 3 shows the relevant regions of the negative ion FAB mass spectrum of the ribonucleotide AAAAAA, the spectrum (a) is that of

DEOXYRIBONUCLEOTIDE GAAGCTTC
NEG. FAB-MS

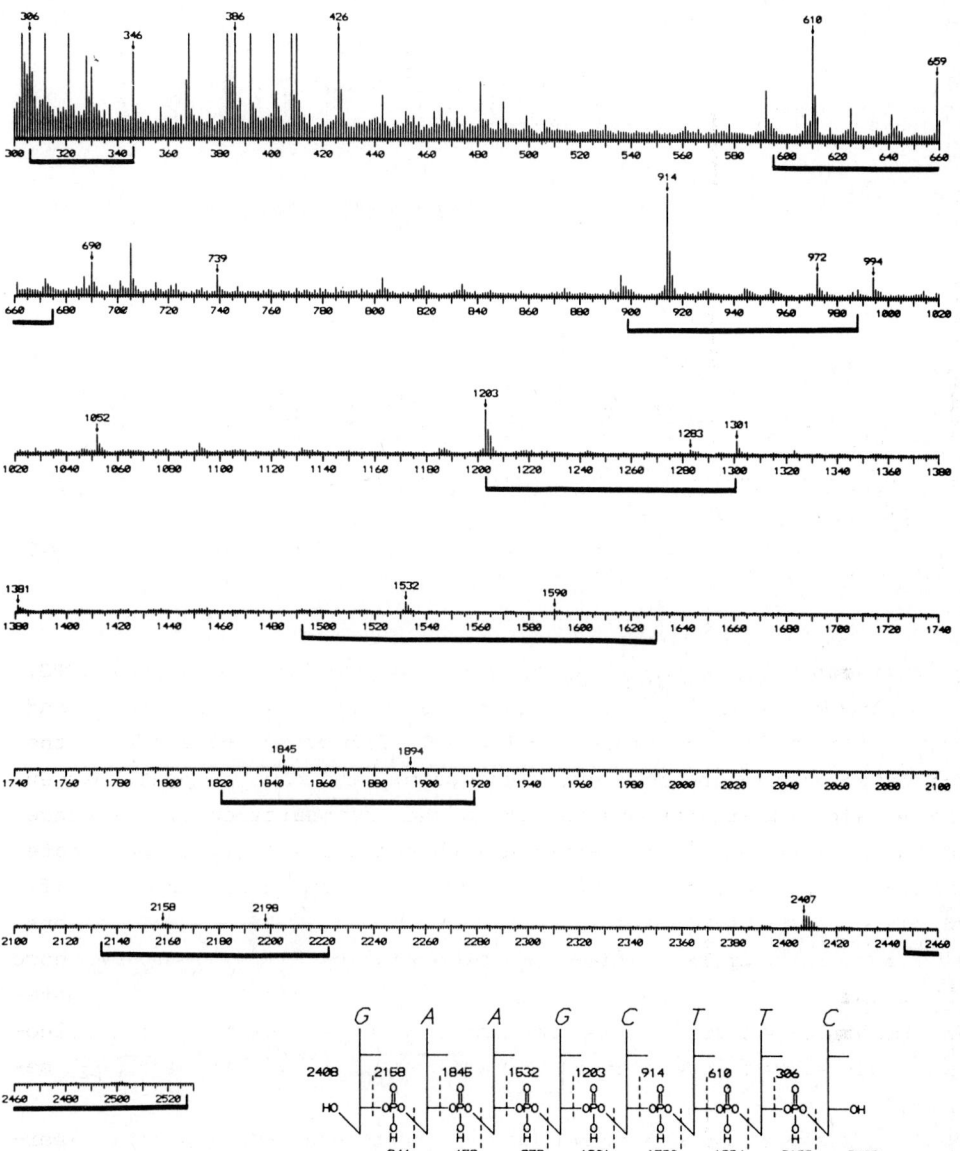

Fig. 2 Negative ion FAB mass spectrum of d(GAAGCTTC). The underlined regions indicate the sequence ion regions.

the sodium salt while spectrum (b) is that of the triethylammonium salt. The spectrum of the sodium salt is complicated by the presence of sodium containing sequence ions. The fourth and the fifth sodium-free sequence ions could not be registered. No such complication is observed for the triethylammonium salt.

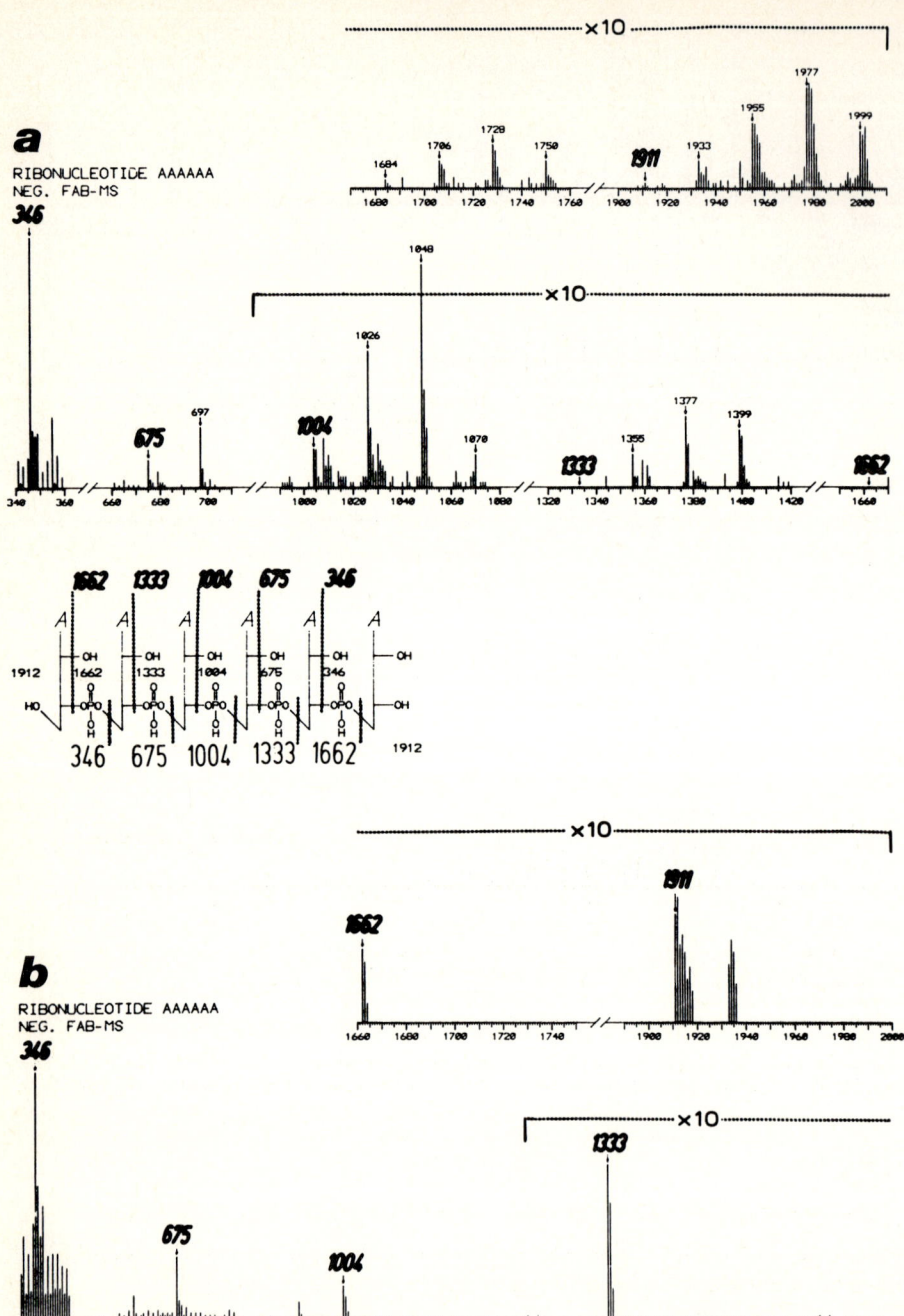

Fig. 3 Sequence ion regions of the negative ion FAB mass spectra of AAAAAA. Top sodium salt; bottom triethylammonium salt.

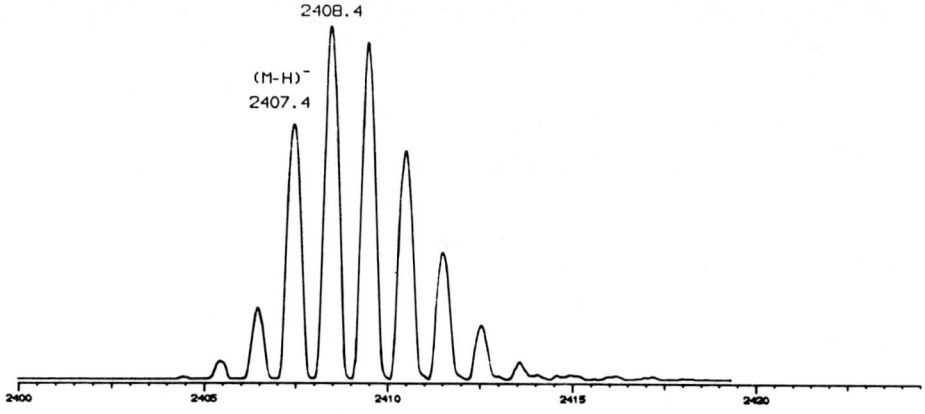

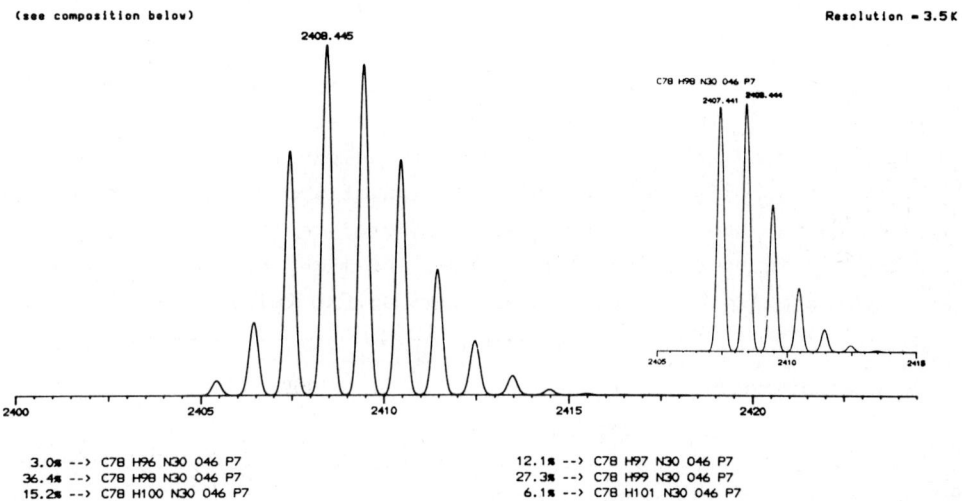

```
3.0%  --> C78 H96  N30 O46 P7        12.1% --> C78 H97  N30 O46 P7
36.4% --> C78 H98  N30 O46 P7        27.3% --> C78 H99  N30 O46 P7
15.2% --> C78 H100 N30 O46 P7         6.1% --> C78 H101 N30 O46 P7
```

<u>Fig. 4</u>   Molecular ion region of the negative ion FAB mass spectrum of d((ACTCGATG). Upper trace experimental result; lower trace simulated results.

As reported previously for free oligodeoxyribonucleotides, all sequence ions were seen without sodium satellites, only the molecular ions have a small sodium-containing partner. Thus we would expect to obtain sequence information of ribonucleotides similar to that of oligodeoxyribonucleotides.

Fig. 4 shows in the upper trace only the deprotonated molecular ion region of the oligodeoxyribonucleotide ACTCGATG. In the lower trace the simulated deprotonated molecular ion cluster is diplayed on the righthand side. The measured result is quite different from this simple simulation. The assumption that only addition or cleavage of hydrogen atoms may occur in the complex molecular ion cluster, is justifiable. The displayed simulation in the lower trace is a mixture analysis of: deprotonated molecular ion 36 %, 1 H less 12 %, 2 H less 3 %; negatively charged molecular ion 27 %, 1 H more 15 %, 2 H more 6 %.

The addition or cleavage of hydrogens on the molecular ion isotope cluster is caused by the liquid matrix glycerol. This effect is less intense for the sequence ions and does not influence the sequence determination.

3 Conclusion

Negative ion FAB-MS is a well-proven analytical method for characterizing synthetic DNA-fragments which are used in genetic engineering. The method allows at present the determination of the molecular ion of protected and free oligonucleotides up to eight and 13 base units respectively, and sequence analysis up to eight and 10 base units.

For modified oligonucleotides, particularly for those where the phosphate group has been altered and rendered resistant to nucleases, this method can be successfully applied and might be the only method of choice, as all classical methods have so far failed [13].

5 Acknowledgements

I would like to thank H. Blöcker and R. Frank of the GBF DNA Synthesis group for the oligonucleotides.

References

1. H. Budzikiewicz and M. Linscheid: Biomed. Mass Spectrom., 4, 103 (1977)

2. H. Budzikiewicz and M. Linscheid: Adv. Mass Spectrom., 7, 1500 (1978)

3. C. J. McNeal, K. K. Ogilvie, N. Y. Theriault and M. J. Nemer: J. Am. Chem. Soc., 104, 972 (1982)

4. C. J. McNeal, K. K. Ogilvie, N. Y. Theriault and M. J. Nemer: J. Am. Chem. Soc., 104, 976 (1982)

5. A. Benninghoven, L. Grotjahn et al: unpublished results.

6. W. Ens, K. G. Standing, J. B. Westmore, K. K. Ogilvie and M. J. Nemer, Anal. Chem., 54, 960 (1982)

7. R. Beavis, W. Ens, M. J. Nemer, K. K. Ogilvie, K. G. Standing and J. B. Westmore, Int. J. Mass Spectrom. Ion Phys. $\underline{46}$, 475 (1983)

8. G. Sindona, N. Ucella and K. J. Weclawek: J. Chem. Res., $\underline{184}$ (1982)

9. L. Grotjahn, R. Frank and H. Blöcker: Nucl. Acids Res., $\underline{10}$, 4671 (1982)

10. M. Panico, G. Sindona and N. Ucella: J. Am. Chem. Soc., $\underline{105}$, 5607 (1983)

11. L. Grotjahn, R. Frank and H. Blöcker: Int. J. Mass Spectrom. Ion Phys., $\underline{46}$, 439 (1983)

12. I. Tazawa, Y. Inoue, S. Seki and M. Kambara: Nucl. Acids Res. $\underline{12}$, 205 (1983)

13. B. A. Nonolly, B. V. Potter, F. Eckstein, A. Pingoud and L. Grotjahn: Biochemistry, $\underline{23}$, 3443 (1984)

14. L. Grotjahn, R. Frank, G. Heisterberg-Moutsis and H. Blöcker: Tetrahedron Lett., $\underline{25}$, 5373 (1984)

Some Experiments on the Production of Ions in Soft Ionisation Mass Spectrometry

D. Renner and G. Spiteller

Institut für Organische Chemie I der Universität Bayreuth,
Universitätsstrasse 30, Postfach 30 08, D-8580 Bayreuth, F. R. G.

Although "soft ionisation" methods are highly recommended to be suitable for accurate molecular weight determination and the structure elucidation of peptides (1) and other polar organic compounds (2) there are many difficulties encountered in practice (3), resulting from the lack of sequence-specific ions. Sometimes such ions are present, sometimes not, why? As recently shown (4), derivatisation may enhance the tendency to produce fragment-specific ions. This is a strong indication that we should be able to direct the production of ions in a desired way if we would know more about the mechanism of fragment production. Peptides are of rather complicated composition, containing basic, acidic and neutral functions within one molecule. This makes investigations about the fragmentation mechanism difficult. We therefore started with the investigation of molecules with only one functional group.

Although the expression "soft ionisation" implies that the investigated molecules are ionized by bombarding particles, there was accumulated in the past a lot of evidence that the "best spectra" are

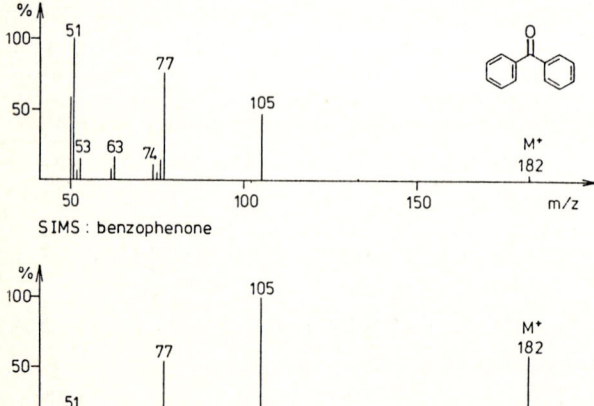

Fig. 1:
a) SIMS: benzophenone
b) EIMS: benzophenone

obtained if there are already ions present in the matrix (5-7). Thus our first aim was to learn to what extent ionisation is achieved by the bombarding particles.

Benzophenone, not able to undergo self-ionisation by protonation, was investigated without matrix. A spectrum (Fig. 1) was obtained showing a M^+ ion and all the fragments typical for its electron impact spectrum. Similar results were obtained with other liquids, but the ion yield was in all cases extremely low. We must conclude that bombarding particles do not ionize a compound in a "soft" way, many of the M^+ ions have enough energy to be cleaved. The higher yield of fragments in LSIMS compared to EI indicates that the M^+ are even "hotter" than those produced by EI.

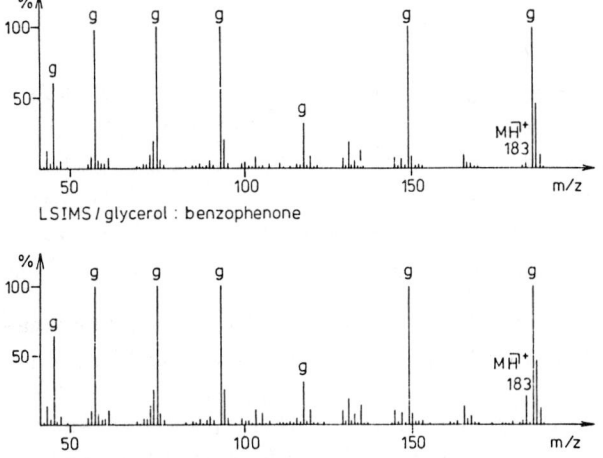

Fig. 2:
a) LSIMS: benzophenone, matrix: glycerol (g)
b) LSIMS: benzophenone after addition of acid

If the spectrum of benzophenone (Fig. 2) was run in glycerol, only MH^+ ions were obtained, buried under the glycerol ions. Addition of HCl improved the intensity for a factor of 4-5.

Dimethylaniline, a basic compound, does also show in glycerol practically no ions. But if we add HCl the intensity of the MH^+ is increased for a factor of at least 100, probably 1000, reflecting the increased ion concentration in solution (Fig. 3).

Thus we can conclude that "soft ionisation" spectra are actually the result of ion desorption. The bombarding particles have only the task to sputter the sample.

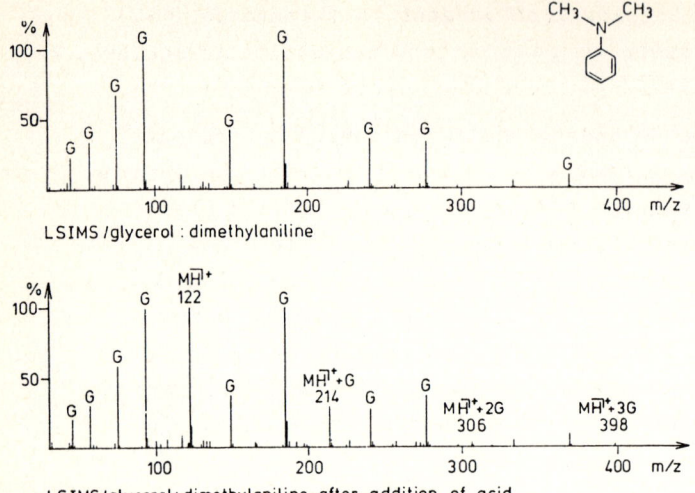

Fig. 3: a) LSIMS: dimethylaniline, matrix: glycerol
b) LSIMS: dimethylaniline after addition of acid

Fig. 4: LSIMS: N-methylbenzanilide, matrix: glycerol

The spectrum of N-methylbenzanilide in glycerol (Fig. 4) taken as an example for a simple amide showed also MH^+ ions. Obviously the proton was either added at the nitrogen or the carbonyl-oxygen.

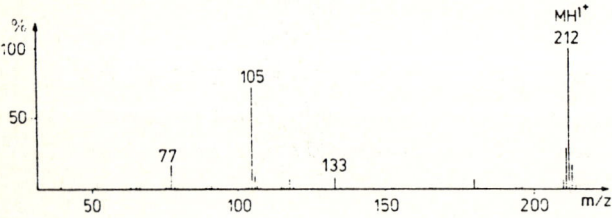

A particle in which the amide nitrogen is protonated needs a very low amount of energy for cleavage of the amide bond, since a stable amine-molecule and a stable benzoyl cation can be produced. Consequently we observe in this case the benzoyl cation and derived from this an ion produced by loss of CO. Addition of HCl does not alter the spectrum drastically, in contrast to amines where the ion yield is much improved.

Since a benzyl cation is also rather stable, we expected that
benzylamines would fragment similarly to benzoylderivatives. To
prove this assumption, N-methylbenzalinide was reduced with
LiAlH$_4$ to the benzyl derivative.

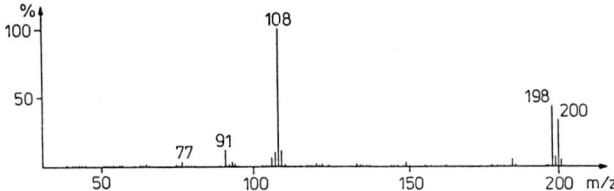

Fig. 5: LSIMS: N-benzyl-N-methylaniline, matrix: glycerol/HCl

Its liquid SIMS spectrum in glycerol after addition of HCl (Fig. 5)
showed the expected fragments of mass 91 but only in low yield. In
addition to the MH$^+$ ion of mass 198 we observed equally intense

ions of mass 108 and 200. Linked scan experiments proved that the
ions of mass 108 are derived from those of mass 200 by loss of
glycerol. The ions of mass 108 correspond to protonated N-methyl-
aniline. The cluster ions of mass 200 must be produced in a complicated reaction, requiring the transfer of two hydrogens from the
matrix surrounding the molecule. Since protonated amines are main
fragments in "soft ionisation" spectra of peptides, these ions
should be produced in similar way.

Peptides occur mainly as zwitter ions. Since their liquid SIMS spectra show MH^+ ions, one proton must be added:

$$H_3\overset{+}{N}\sim\sim COO^- \longrightarrow H_3\overset{+}{N}\sim\sim COOH$$
$$H_3\overset{+}{N}\sim\sim COO^- \longrightarrow H_2N\sim\sim COO^-$$

The matrix molecules contain only OH hydrogens of low acidity. The acidity of a peptide is much greater than of glycerol. Therefore we must assume that most of the MH^+ ions are produced by transfer of a proton from one peptide molecule to the next, diminishing the ion yield of MH^+ at the best to 50%.

Therefore it is not astonishing that addition of acid increases the ion yield. Besides protonation at the basic nitrogen, addition of

protons seems also to be possible in a very small amount at peptide bonds. We have to assume that even dications are formed. In these ions as shown with amides an easy cleavage of the peptide bond must be expected. Nevertheless, even in the presence of acid only very few peptide molecules will be present in this form.

Having in mind that already charged particles show much more intense spectra than neutral ones, we prepared a charged peptide-derivative by reacting quaternary sulfonylchloride with a peptide:

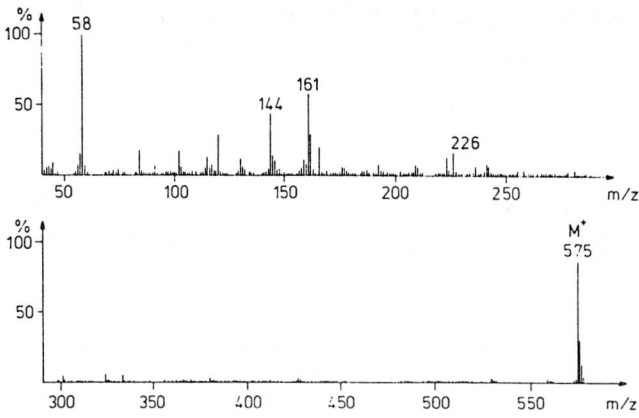

Fig. 6: LSIMS: Glu-Gly-Phe derivatised with chlorosulfonyl-N,N-dimethyl-1,2,3,4-tetrahydroisochinolinium chloride, matrix: glycerol (subtracted)

The spectrum (Fig. 6) showed very intense M^+ ions, but any sequence-specific ions were missing (indicated fragments originate from the protecting group). Thus we must conclude that the location of charge triggers fragmentation reactions. In agreement with this assumption, the LSIMS spectra of free peptides show often abundant imminium-ions of the N-terminal amino acid which is preferentially protonated at its basic nitrogen, but only a few structure-specific fragments resulting from the cleavage of peptide bonds.

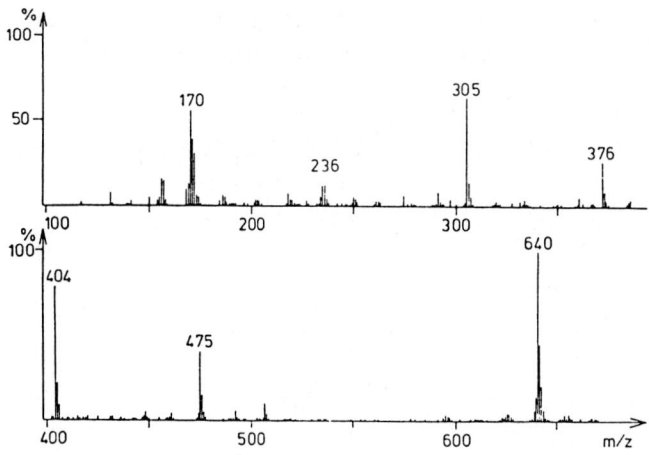

Fig. 7: LSIMS: Dns-Val-Ala-Ala-Phe, MH^+ = m/z 640, matrix: glycerol (subtracted)

Therefore we expected if we transform the basic nitrogen in a neutral amide by acylation we should have in the molecule only amide bonds, able to undergo structure-specific fragmentation. Actually a literature research revealed that acetylated peptides usually showed sequence-specific ions (3).

Following this hint we prepared dansyl derivatives of peptides. They showed intense acyl ions produced by cleavage of the amide bonds.

An even better recognition of the sequence-specific ions is achieved if a peptide is reacted with 2-bromo-5-(dimethylamino)-benzenesulfonylchloride (Bdbs-Cl), due to the bromine isotopes, although an unexpected side reaction by exchange of bromine to hydrogen occurred.

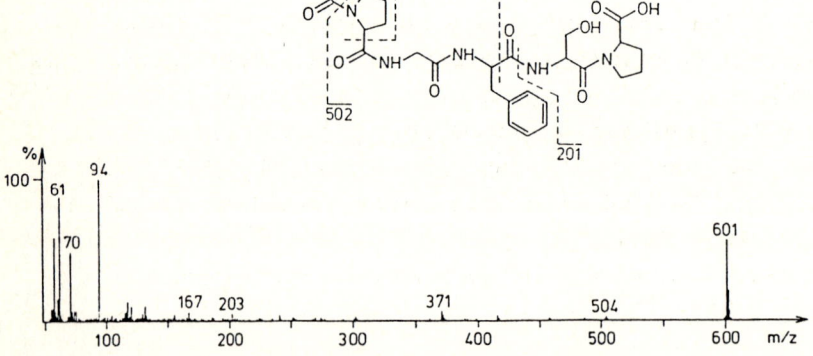

Fig. 8: LSIMS: Pro-Pro-Gly-Phe-Ser-Pro, MH$^+$= m/z 601, matrix: glycerol (subtracted)

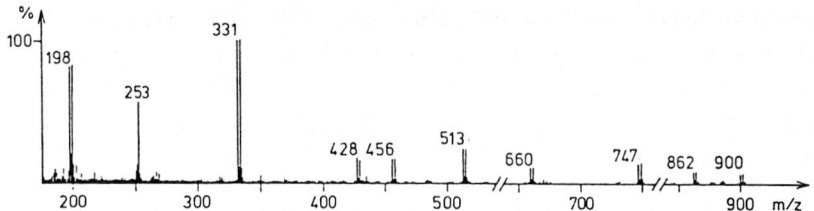

Fig. 9: LSIMS: Bdbs-Pro-Pro-Gly-Phe-Ser-Pro, MH^+ = m/z 862, MK^+ = m/z 900, matrix: glycerol (subtracted)

We hope that our investigation of model compounds stimulate further investigations about fragmentation reactions of ions produced by "soft ionisation" techniques, which actually are desorption methods.

We thank Deutsche Forschungsgemeinschaft, Robert-Pfleger-Stiftung in Bamberg and the Fonds der Chemischen Industrie for financial support.

Literature

(1) A. Dell, H.R. Morris, Biochem. Biophys. Res. Commun., **106**, 1456 (1982)
(2) A. Dell, G.W. Taylort, Mass Spectrom. Rev., **3**, 357 (1984)
(3) P. Roepstorff, P. Hojrup, J. Moller, Biomed. Mass Spectrom., **12**, 181 (1985)
(4) D. Renner, G. Spiteller, Angew. Chem., **97**, 408 (1985)
(5) T.M. Ryan, R.J. Day, R.G. Cooks, Anal. Chem., **52**, 2054 (1980)
(6) D.H. Williams, C. Bradley, G. Bojesen, S. Santikarn, L.C.E. Taylor, J. Am. Chem. Soc., **103**, 5700 (1981)
(7) E. Schneider, K. Levsen, P Dähling, F.W. Röllgen, Fresenius Z. Anal. Chem., **316**, 277 (1983)

Decompositions Occurring Remote from the Charge Site: A New Class of Fragmentation of FAB-Desorbed Ions

K.B. Tomer and M.L. Gross

Midwest Center for Mass Spectrometry, University of Nebraska,
Lincoln, NE 68588, USA

1. Introduction

For the past several years we have been involved in a research program to develop the combined techniques of fast atom bombardment and tandem mass spectrometry for structure determination [1-18]. This research has been performed on a Kratos MS-50 triple sector mass spectrometer which consists of a high resolution MS-50 as MS-I coupled to an electrostatic analyzer serving as MS-II [19]. The ion selected with MS-I has a translational energy of 8KeV and the collision process transfers up to one hundred electron volts into internal energy. This makes accessible reaction channels which are not readily found for ions with the usual range of internal energies. As a result of the high energy, a hitherto unrecognized class of reactions has been discovered. These reactions are most readily observed for closed-shell ions and occur remote from the charged site, in contrast to the common charge-induced reactions. In this paper, the experimental evidence for the nature of the reaction and the extent of its occurrence will be discussed.

2. Experimental Evidence

The first evidence which led to recognition of these reactions arose during a set of experiments of the behavior upon collisional activation of FAB-desorbed unsaturated fatty acid anions [1]. The CAD spectrum (Fig. 1) of a typical unsaturated fatty acid is dominated by a series of high mass ions formed by losses of (14n+2) amu. The gap in the pattern is due to the presence of the double bond, which suggests that the major fragmentations are occurring at the end of the alkane chain remote from the charge site.

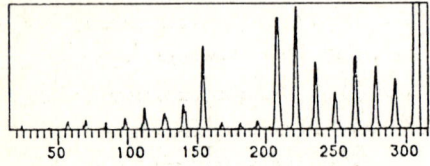

Fig. 1. CAD spectrum of (M-H) anion of 11-Eicosenoic acid.

The 14 dalton gap between ions suggests that the losses observed are of C_nH_{2n+2}. Validation of this assumption was obtained by investigating saturated fatty acids and deuterium labeled analogs. The initial loss of 16 amu was shown to be CH_4 loss in the CAD spectrum of the (M-H)$^-$ ion of palmitic acid by the observation of 19 amu loss (CD_3H) from 16,16,16-d_3-palmitic acid. Only the mass of the (M-H)$^-$ ion is different in the spectra of the labeled and unlabeled compounds. Further validation is found in the CAD spectra of the $(M-H)^-$ ions of other labeled carboxylic acids including 2,3-d_2-octanoic acid, 9,10-d_2-stearic acid, 7,7,8,8-d_4-palmitic acid, 9,10,12,13-d_4-stearic acid, 9,10,12,13,15,16-d_6-stearic acid, 5,6,7,8,11,12,14,15-d_8-eicosanoic acid and 4,5,7,8,10,11,13,14,16,17,19,20-d_{12}-docosahexanoic acid [2,6].

Mechanisms to accommodate these results are shown in eqn (1) and (2). A six-centered reaction, eqn (1), is preferred over the four-centered reaction, eqn (2), for these reasons: 1) the process in eqn 1 is allowed by orbital symmetry considerations while eqn (2) is not. The process in eqn 2 should also give rise to saturated carboxylate anions and unsaturated alkane neutrals which are not observed, and 3) fatty acid methyl esters have been shown by Sun et al. to undergo pyrolysis to form alkenes and unsaturated fatty acid esters [20].

$$CH_3(CH_2)_nCH_2\overset{H}{\underset{CH_2}{\diagdown}}CH_2-(CH_2)_mCOO^- \longrightarrow CH_3(CH_2)_nCH_2CH_3 \quad (1)$$
$$+ CH_2=CH(CH_2)_mCOO^-$$

$$CH_3(CH_2)_nCH\overset{H\ H}{\underset{CH_2-CH_2}{\diagdown\diagup}}CH-(CH_2)_mCOO^- \longrightarrow H_2 + CH_3(CH_2)_nCH_2=CH_2 \quad (2)$$
$$+ CH_2=CH(CH_2)_mCOO^-$$

In a study of the energetics of remote-site fragmentation in phosphonium ions (see below), the activation energy was observed to be greater than 100 eV [15]. Fraley and Lawrence have investigated remote charge site fragmentation by using variable collision energies and observed that minimal C_nH_{2n+2} loss was observed below ca. 200eV [18]. Thus, these reactions have a relatively high activation energy. Reactions that clearly involve the charged group, such as water loss, were still observed at low collision energies (<100eV).

These data corroborate our view that this reaction type is a significant departure from the widely accepted idea that fragmentations of gas-phase ions is initiated by a charged-site or radical-site. The evidence points to a thermal reaction occurring within the molecule, the increased charge density near the negative charge-bearing moiety mitigating against thermal bond cleavage in its proximity.

3. General Nature of the Reaction

Remote charge site reactions are most readily observed for closed-shell negative ions. If, however, the mechanism is general, remote site reactions should be observed in the CAD spectra of diverse compound types typically in the presence of closed-shell positive ions, e.g. quaternary ammonium ions, other closed-shell negative ions and in systems where the charge is highly localized. The following is a listing of compound types which undergo remote site decomposition reactions.

3.1. Modified Fatty Acids

In addition to the saturated and unsaturated fatty acids, the CAD spectra of the (M-H) anions of a series of substituted fatty acids, including epoxy acids, hydroxy acids, cyclopropane and cyclopropene containing acids and branched-chain acids, have been found to be dominated by remote-site fragmentations [22]. The fragmentations are characteristic of the substituent, and provide structural information as to the nature and location of the substituent. The CAD spectrum of the (M-H) anion of vaccenic epoxide is presented as an example (Fig. 2). Other modified fatty acids such as fluorinated fatty acid sulfates and sulfonates also exhibit remote site fragmentation [14].

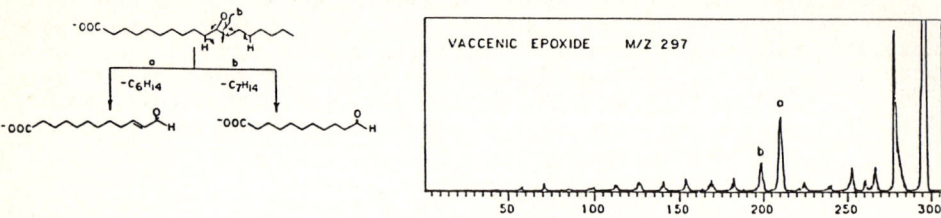

Fig. 2. CAD spectrum and fragmentation mechanism of the (M-H) anion of vaccenic epoxide.

Under positive ion conditions, the CAD spectra of the $(M+H)^+$ ions of fatty acids are dominated by charge-induced low mass fragment ions. Under conditions in which the positive charge is highly localized, however, such as in $(RCO_2Li_2)^+$ ions, the CAD spectra now show significant remote-site fragmentation in addition to the charge-induced fragmentation [22].

3.2. Vitamin E

Vitamin E contains an isoprenoid side chain. The CAD spectrum of the (M-H) anion contains remote-site fragmentations which clearly locate the site of branching (Fig. 3).

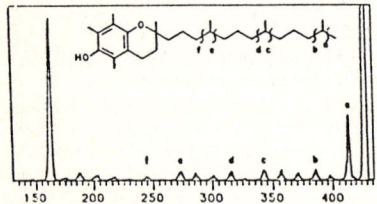

Fig. 3. CAD spectrum & fragmentation mechanism of the (M-H) anion of Vitamin E.

3.3. Ammonium Ions

Quaternary ammonium ions, as closed-shell positive ions, are expected to exhibit remote-site fragmentations. The CAD spectrum of the dimethyldidecylammonium cation (Fig. 4) contains abundant ions resulting from C_nH_{2n+2} loss. Other ammonium ions such as

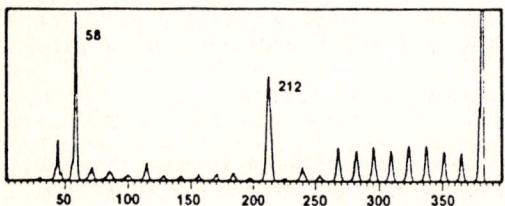

Fig. 4. CAD spectrum of dimethyl didodecylammonium cation.

alkylpyridinium ions and ethoxylated quaternary amines also exhibit remote-site fragmentation upon collisional activation [4].

3.4. Phosphonium Ions

The CAD spectra of alkyltriphenylphosphonium ions show both proximate and remote site fragmentations (Fig. 5). Structural information as to the nature of the alkyl chain is readily observed. The ratio of remote to proximate fragmentation increases from 0 to 135 as the collision energy is increased from 100 to 8000 eV [15].

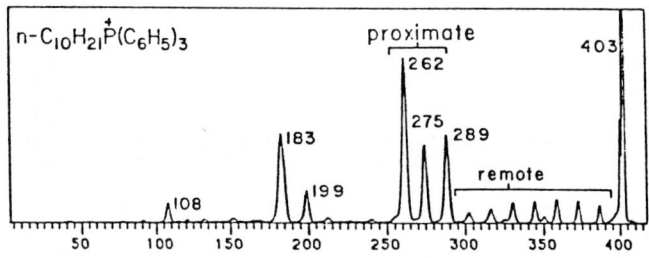

Fig. 5. CAD spectrum of decyltriphenylphosphonium cation.

3.5. Bile Salts

The major fragment ions, save for H_2O expulsion, formed by collisionally activating the (M-H) anions of the taurine and glycine conjugates of bile acids are due to remote site ring cleavages (eqn 3). Cleavage A is typically the favored cleavage and is the most remote from the charge site [16]. The concept of remote-site fragmentation was also found to be crucial to an understanding of fragmentation differences observed in conjugated bile acid sulfates.

3.6. Nucleotides

A major collisionally activated decomposition of the (M-H) anion of nucleotides, modified nucleotides and di- and trinucleotides arises by expulsion of a neutral base. This involves hydrogen transfer from the sugar moiety to the base [17,22]. The charged site on these molecules is presumably the phosphate group, which is remote from the site of the reaction (eqn 4).

3.7. Glutamylcarboxyphenylhydrazine

N^2-(γ-L(+)glutamyl)-4-carboxyphenylhydrazine (GLPH), a potential carcinogen, has been identified in commercially cultivated mushrooms [18]. The structure proof of the compound relied heavily on the CAD spectra obtained for its $(M+H)^+$ and $(M-H)^-$ ions. Although the molecule contains many sites capable of bearing the charge, the CAD spectrum of the (M-H) anion was best interpreted on the basis of remote-site fragmentation (eqn 5 and 6).

$$\text{-O-C(=O)-C}_6\text{H}_4\text{-NH-NH-CO-CH}_2\text{-CH}_2\text{-CH(NH}_2\text{)-COOH} \quad (5)$$

$$\text{HO-C(=O)-C}_6\text{H}_4\text{-NH-NH-CO-CH}_2\text{-CH}_2\text{-CH(NH}_2\text{)-COO}^- \quad (6)$$

3.8. Steroid Glucuronides

The CAD spectra of the (M-H) anions of 5β-androstan-3α-ol-11,17-dione glucuronide and 5α-androstan-3α-ol-11,17-dione glucuronide are different in the relative abundance of the ion from $C_6H_8O_6$ loss (eqn 7) (5b>5a) [22,23]. This reaction occurs remote from any probable charge-bearing moiety. Thus, remote charge-site fragmentations show potential as a stereochemical probe.

4. Conclusion

The CAD of closed-shell ions has led to the observation of a new class of reactions, the remote charge-site fragmentation. This reaction is most easily envisioned as a thermal reaction of the ion, not involving nor initiated by the charged site. Significant fragmentation processes in both positive and negative ions of a wide variety of chemical species can most easily be rationalized by invoking this process.

5. Acknowledgements

The generous support of the U.S. National Science Foundation (Grants CHE-8211164 and CHE-8320388) is gratefully acknowledged.

6. References

1. Tomer, K.B.; Crow, F.W.; Gross, M.L., J. Am. Chem. Soc. 105, 5487 (1983).
2. Jensen, N.J.; Tomer, K.B. and Gross, M.L., J. Am. Chem. Soc. 107, 1863 (1985).
3. Lyon, P.A.; Stebbings, W.L.; Crow, F.W.; Tomer, K.B.; Lippstreu, D.L.; Gross, M.L., Anal. Chem. 56, 8 (1984).
4. Lyon, P.A.; Crow, F.W.; Tomer, K.B.; Gross, M.L., Anal. Chem. 56, 2278 (1984).
5. Jensen, N.J.; Tomer, K.B.; Gross, M.L.; Lyon, P.A., Proceedings of the FAB/SIMS Symposium (in press).
6. Jensen, N.; Tomer, K.; Gross, M., Anal. Chem., in press (1985).

7. Gross, M.L.; McCrery, D.; Crow, F.; Tomer, K.B.; Pope, M.R.; Ciufetti, L.M.; Knoche, H.W.; Daly, J.M.; Dunkle, L.D., Tet. Letts. 23, 5381 (1982).
8. Tomer, K.B.; Crow, F.W.; Gross, M.L.; Kopple, K.D., Anal. Chem. 56, 880 (1984).
9. Bodley, J.W.; Upham, R.; Crow, F.W.; Tomer, K.B.; Gross, M.L., Arch. Biochem. Biophys. 230, 590 (1984).
10. Crow, F.W.; Tomer, K.B.; Gross, M.L.; McCloskey, J.A.; Bergstrom, D.E., Anal. Biochem. 139, 243 (1984).
11. Tomer, K.B.; Crow, F.W.; Knoche, H.W.; Gross, M.L., Anal. Chem. 55, 1033 (1983).
12. Kim, S.-D.; Knoche, H.W.; Dunkle, L.D.; McCrery, D.A.; Tomer, K.B., Tet. Letts. 26, 969 (1985).
13. Lippstreu-Fisher, D.; Gross, M., Anal. Chem. 57, 1174 (1985).
14. Lyon, P.A.; Tomer, K.B.; Gross, M.L., Anal. Chem. in press.
15. McCrery, D.; Peake, D.; Gross, M., Anal. Chem. 57 1181 (1985).
16. Tomer, K.B.; Jensen, N.J.; Gross, M.L.; Whitney, J.D., Biomed. Mass Spectrom. submitted.
17. Tomer, K.; Gross, M.; Deinzer, M., J. Am. Chem. Soc. sub.
18. Chauhan, Y.; Nagel, D.; Gross, M.; Cerny, R.; Toth, B., J. Agr. Food Chem. in press.
19. Gross, M.L.; Chess, E.K.; Lyon, P.A.; Crow, F.W.; Evans, S.; Tudge, H., Int. J. Mass Spectrom. Ion Phys. 42, 243 (1982).
20. Sun, K.; Hayes, H.; Holman, R., Org. Mass Spec. 3, 1035 (1970).
21. Fraley, D.F.; Lawrence, D.L., Am. Soc. Mass Spect. Proceedings 33rd Annual Conference, San Diego, 1985.
22. Unpublished reports.
23. Cheng, M.T.; Barbalas, M.P.; Pegues, R.F.; McLafferty, F.W., J. Am. Chem. Soc. 105, 1510 (1983).

Part IV

Laser-Induced Ion Formation

Laser and Plasma Desorption: Matrices and Metastables in Time-of-Flight Mass Spectrometry

R.J. Cotter, J. Honovich, J. Olthoff, P. Demirev, and M. Alaim

Department of Pharmacology, The Johns Hopkins University, Baltimore, MD 21205, USA

1. Introduction

Pulsed laser desorption and plasma desorption have in common the fact that they are used on solid samples (rather than liquid substrates as in fast atom bombardment) and that they are most appropriately used with the time-of-flight analyzers. In the course of our application of both of these techniques to several analytical problems, our attention has been focussed not only on the differences in the ionization processes, but also on the importance of sample preparation, the properties of the TOF analyzer, and the detection system. Most interestingly we find that as in FAB, the choice of chemical matrices for the sample can have a profound effect on the spectra, and that the two TOF analyzers used record metastable decompositions in very different ways.

2. Experimental

Plasma desorption mass spectra were recorded on a BIO-ION Nordic mass spectrometer with a 15 cm flight tube and 20 kV accelerating voltage. Ions recorded on a 1ns/channel multistop time-to-digital converter (TDC) were processed in an Apple IIe microcomputer. Laser desorption mass spectra were recorded on an instrument designed in our laboratory [1] and consisting of a Tachisto 215G carbon dioxide laser and a CVC 2000 time-of-flight analyzer, with a 1 m flight tube and 3kV accelerating voltage. The ion source region is at ground, so ions are produced in a field-free region, and then extracted by pulsing the first two stages of a multiple accelerating field region prior to entering the drift tube (analyzer). Data are recorded on a LeCroy 3500SA signal averaging system using a 100MHz waveform recorder. The instruments are compared in Figure 1.

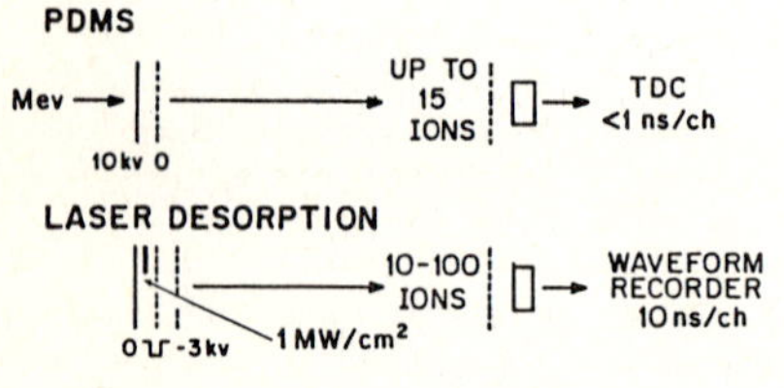

Fig. 1

3. Results and Discussion

3.1. Phosphorylated Lipo-oligosaccharides

Lipo-oligosaccharides consist of two amino sugar groups to which are attached several O-acyl and N-acyl linked fatty acid chains, generally at the 2 and 3 positions. They are isolated from gram-negative bacteria and are generally phosphorylated on the 6 position of the distal sugar, which makes them negatively charged. Negative ion FAB spectra produce stable negative molecular ions with little fragmentation, while positive ion spectra with more information can be

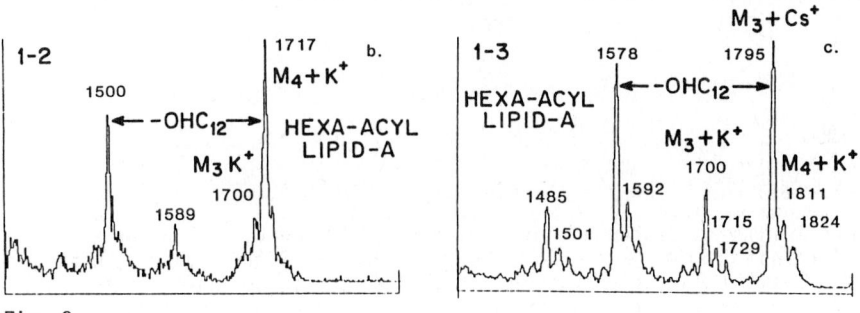

dimethyl monophosphoryl hexa-acyl LIPID-A

M_3 = 2 OHC$_{12}$, 2 OHC$_{14}$, 2 n-C$_{12}$ 1661
M_4 = 3 OHC$_{12}$, 2 OHC$_{14}$, 1 n-C$_{12}$ 1677

obtained using thioglycerol which, among other things, acts as an acid to protonate the negatively charged phosphate groups. These compounds have been studied extensively by LINDNER [2] using laser desorption of the de-phosphorylated lipid A. In the past we have reported the laser desorption spectrum of mono phosphoryl lipid A using NH$_4$Cl as a solid acidic matrix [3]. In more recent work we have used the dimethyl phosphoryl derivatives of lipid A which provides greater ion signal for the molecular ion. The spectra in Figure 2 shows the molecular ion regions of two lipid-A components, seperated by HPLC.

Fig. 2

3.2 Metastable Decomposition of Ions

The plasma desorption mass spectrometer uses a static field analyzer, while the laser desorption instrument uses a dynamic (gated) analyzer in which the accelerating field is imposed some time after the ionization event. In the static field analyzer most of the fragment ions (from ions decomposing in less than 10^{-8} seconds) are formed in the accelerating field. If they are formed promptly they will appear as fragment ions in the spectrum, but if they are formed after partial accelerations they will produce the "tailing" and "continuum" commonly observed in plasma desorption spectra [4]. If, on the other hand, the ions are formed in a field-free region and extracted only after decomposition is nearly completed, then the effects of metastable defocussing on the spectra can be minimized or eliminated. Some comparative spectra in Figures 3 and 4 for plasma and laser desorption illustrate this point.

3.3 Organic Matrices in Plasma Desorption

The effect of matrices in FAB mass spectrometry is recognized by almost all who employ this technique, and several laboratories have described their own "magic bullet" [5]. Designing appropriate chemical matrices for solid techniques is not so clear, although MACFARLANE has noted that desorption of molecular ions

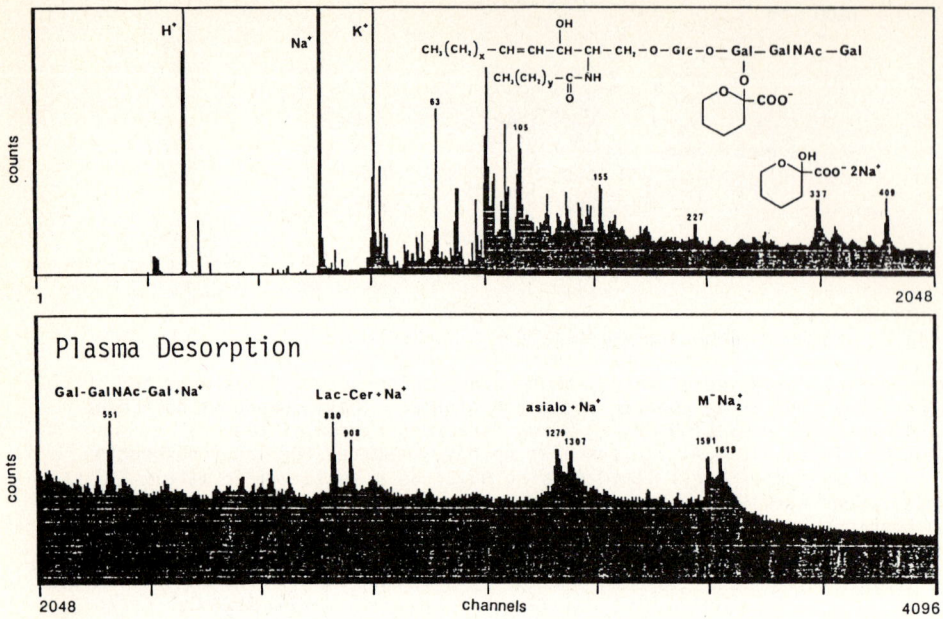

Fig. 3

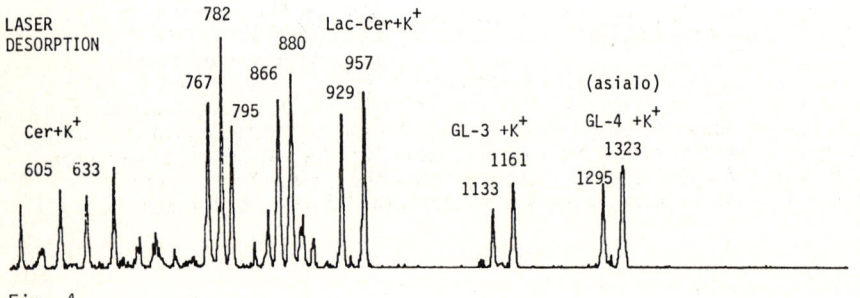

Fig. 4

from monolayers of sample on a clear metal surface is far different from one in which the sample, or substances chemically similar to the sample, provides an insulating layer on the surface [6]. Using this "like dissolves in like" philosopy, we have recorded plasma desorption spectra of several disulfide-containing peptide hormones in a mixture of oxidized and reduced glutathione. In figure 5, the spectra of relaxin, an insulin homolog, recorded with and without glutathione, shows a remarkable difference in appearance with respect to the molecular ion signal-to-noise and the extent of the baseline continuum. The extent to which this matrix effects the mass spectrum can be appreciated in the spectrum of human proinsulin, shown in Figure 6. From this spectrum the molecular ion mass can be readily determined without the extensive, computer assisted correction for the background continuum which has been necessary previously [7].

3.4 Post Accelerating Detection

We have noted that the laser desorption technique has not at this time achieved the spectacularly high mass measurements of fast atom bombardment in the 5,000

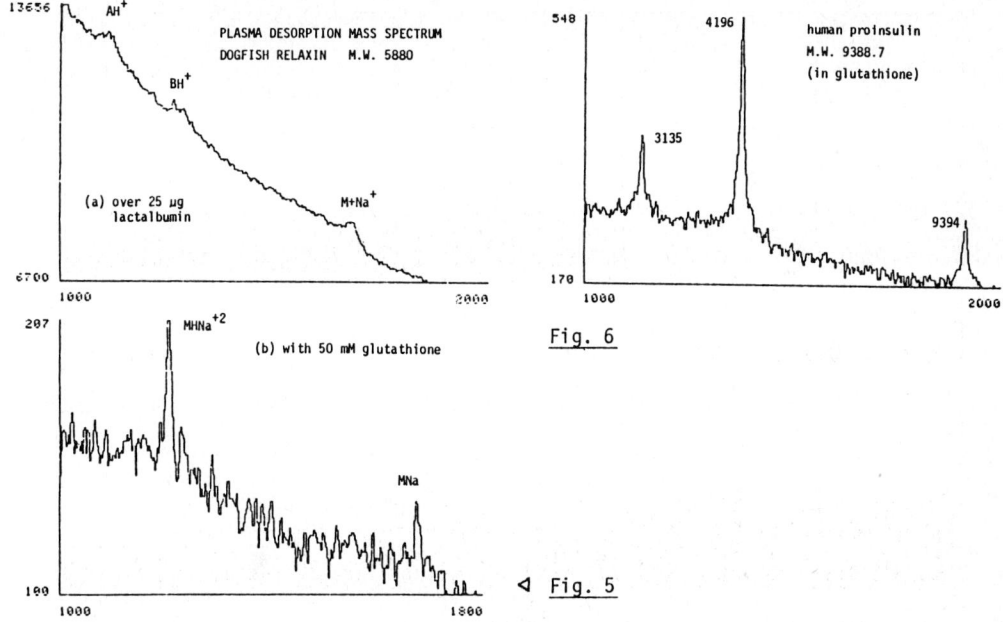

Fig. 6

Fig. 5

to 15,000 amu range. Some time ago we reported the molecular ion of phosphazine at 3628 amu by laser desorption [8]. We attribute this mass limitation to the particle detector and relatively low accelerating voltages used in laser TOF instruments, rather than to the ionization method; and we are encouraged by WILKENS' measurement of oligomers to 7000 amu on an FTMS instrument which does not employ particle detection [9]. Recently we have constructed a dual channelplate detector which provides an additional 5kV acceleration to the 3keV ions in the laser TOF mass spectrometer. Our initial experiments have been concerned with the effect of the post acceleration region and the detector itself in the time/mass resolution of spectra obtained in the analog mode.

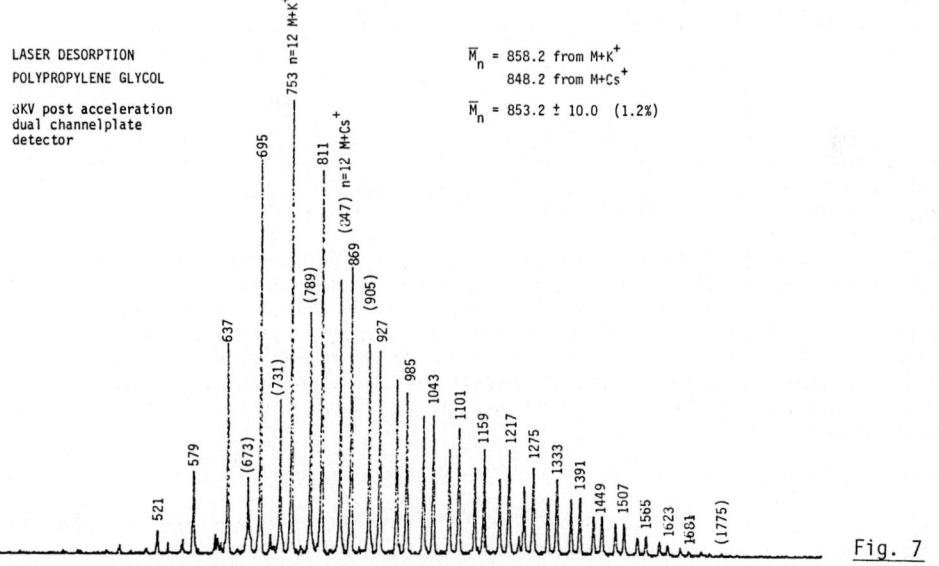

Fig. 7

Figure 7 shows the mass spectrum of a mixture of polypropylene glycol 800 and 1200 oligomers. The resolution appears to be as good or better as that obtained on the magnetic electron myltiplier, and comparison of the average molecular weights from H+K$^+$ and M+Cs$^+$ ions show little mass discrimination.

ACKNOWLEDGEMENTS

The authors acknowledge the contributions to this work by K. Takayama, N. Qureshi, I. Kamensky, H. Fales and C. Fenselau. This work was supported by NSF grants PCM-8209954 and CHE-8410506.

1 R.B. VanBreemen, M. Snow and R.J. Cotter, Int. J. Mass Spectrom. Ion Phys., 49, 35 (1983)
2 U. Seydel and B. Lindner, "Ion Formation from Organic Solids", A. Benninghoven, Ed.; Springer-Verlag, Berlin, 1982, p 240
3 J.-C. Tabet and R.J. Cotter, Anal. Chem., 56, 1662 (1984)
4 B.T. Chait and F.H. Field, Int. J. Mass Spectrom. Ion Proc., 65, 169 (1985)
5 J.L. Witten, M.H. Schaffer, M. O'Shea, J.C. Cook, M.E. Hemling and K.L. Rinehart, Jr., Biochem. Biophys. Res. Commun., 124, 350 (1984)
6 R.D. Macfarlane, Accts. Chem. Res., 15, 268 (1982)
7 B. Sudqvist, I. Kamensky, P. Hakansson, J. Kjellberg, M. Salehpour, S. Widdiyasekera, J. Fohlman, P.A. Peterson and P. Roepstorff, Biomed. Mass Spectrom., 11, 242 (1984)
8 R.J. Cotter and J.-C. Tabet, Int. J. Mass Spectrom. Ion Phys., 53, 151 (1983)
9 C.L. Wilkens, D.A. Weil, C.L.C. Yang and C.F. Ijames, Anal. Chem., 57, 520 (1985)

Evidence for Simultaneous Generation of Ion Pairs in Laser Mass Spectrometry

D.M. Hercules

Department of Chemistry, University of Pittsburgh, Pittsburgh, PA 15260, USA

1. Introduction

Our research group has been concerned with the laser mass spectra of organic compounds; particular emphasis has been on fragmentation processes and the mechanisms which produce organic ions. Frequently, positive and negative ions occur in pairs raising the question of whether ion formation can result from charge exchange reactions producing an ion pair in a single step (1,2). Pair-production could be brought about by intermolecular transfer of atoms, groups, or electrons, or by intramolecular charge separation. The general reactions for these possibilities can be summarized below:

$$2GM \xrightarrow{h\nu} {}^+MG_2 + M^- \tag{1}$$

where G = H or an alkyl group

$$AB \xrightarrow{h\nu} A^+ + B^- \tag{2}$$

The above reactions imply pair-production in a single step, which distinguishes them from indirect processes which can produce the same ions. General examples of the latter are:

$$GM \xrightarrow{h\nu} G^+ \xrightarrow{GM} MG_2^+ \tag{3}$$

$$AB \xrightarrow{h\nu} AB^{+*} \longrightarrow A^+ + B \tag{4}$$

Production of pairs of ions in the laser mass spectra of a variety of systems is a well-documented fact. Formation of $(M + H)^+$ and $(M - H)^-$ ions is common for molecules containing **acidic** and **basic** groups; for example, carboxylic acids, amino acids, and phenols. Also, in both the field desorption and laser mass spectra of zwitterionic compounds, ions such as $(M + CH_3)^+$ and $(M - CH_3)^-$ are observed for the same compound. Similarly, pair production from a single molecule is possible; for example, esters (R_1COOR_2) show R_1CO^+ and OR_2^- peaks in their positive and negative ion laser mass spectra, respectively. Almost all aromatic hydrocarbons show peaks corresponding to $M^{+\cdot}$ in their positive ion spectra, and many show $M^{-\cdot}$ ions in their negative ion spectra. This raises the possibility of intermolecular electron transfer reactions to produce positive and negative ions.

The purpose of the present paper will be to explore evidence for ion pair production in the laser mass spectrometry of organic compounds.

2. Theoretical Considerations

It is difficult to find an exact theoretical framework in which to discuss ion pair formation in organic compounds; this probably reflects the general state of confusion regarding the mechanisms of ion formation in laser mass spectrometry. Intramolecular ion pair formation in the gas phase has been addressed [3]. Based on statistical theory, it was concluded that ion pair formation can represent an important process in gas phase ion formation near the energy threshold. However, as the energy of the system is increased, ion pair formation becomes less important, primarily due to electron detachment of the negative ion. Thus, there is precedent for the concept of ion pair production in laser mass spectrometry, although extrapolation from an intramolecular gas-phase process to an intermolecular solid state process is indeed a giant step.

Additional justification for the concept comes from shock tube studies [4]. Shock initiated thermal dissociation of alkali metal halides produced an inverted distribution of dissociated species, yielding primarily alkali metal and halide ions, rather than dissociated neutral atoms in their ground state. The distribution did not correspond to a thermal distribution of dissociated atoms and ions. Again, although extrapolation from alkali metal halide salts to organic molecules is a rather large one, one must remember that shock wave phenomena are not unimportant for producing ions in laser mass spectra. Thus, again, there appears to be precedent.

We have recently considered [5] the energetics of proton transfer reactions in the laser ionization of amino acids. Professor K. Jordan constructed a correlation diagram for an amino acid dimer (hypothetical) which is shown in Figure 1. Although the curves are approximate, they show the concept of ion pair formation by (in this case) proton transfer. Curve A is a potential energy surface which correlates with two uncharged amino acids at separation. A thermal process will bring about adiabatic separation of the dimer yielding two uncharged molecules. A higher energy process, Curve B, corresponds to the production of one zwitterion and one unionized molecule; Curve C correponds to production of two zwitterions. Neither of these processes will produce ions in the mass spectrum directly. Curve D corresponds to separation of two ions by proton transfer in an amino acid dimer to yield $(M + H)^+$ and $(M - H)^-$.

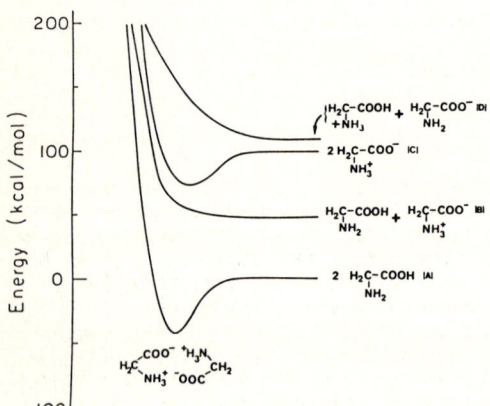

Fig. 1. Correlation diagram for amino acid dimers

The important idea of Figure 1 is that simple thermal separation of an amino acid pair cannot produce $(M + H)^+$ and $(M - H)^-$ ions; higher energy curves are necessary. The possibility for crossover between curves A and D in Fig. 1 is small because of the large energy separation between them. Energetics also indicate that two photons used in the experiment ($h\nu$ = 265 nm) would be sufficient to raise the system to Curve D, producing the ion pair.

The above theoretical considerations indicate that ion pair formation might occur. Now it is important to explore experimental observations to determine if they are consistent with pair production.

3. Evidence from the Amino Acids

The simplest reaction one can imagine for production of quasimolecular ions in the amino acids would be direct proton transfer from one molecule to another. Amino acids exist in the solid as zwitterions; thus, transfer of a single proton could produce an ion pair:

$$\text{R-CH}\begin{array}{c}\text{NH}_3^+\\\text{COO}^-\end{array} + \begin{array}{c}\overline{\text{OOC}}\\\text{HNH}_2^+\end{array}\text{CH-R} \xrightarrow{h\nu} \text{RCH}\begin{array}{c}\text{NH}_3^+\\\text{COOH}\end{array} + \begin{array}{c}\overline{\text{OOC}}\\\text{NH}_2\end{array}\text{CH-R} \quad (5)$$

However, it is also possible that acid base reactions can occur from species produced by laser decomposition of amino acids. A possible sequence of steps might be:

$$HA \xrightarrow{h\nu} H^+ + \text{products} \quad (6)$$

$$HA + H^+ \longrightarrow H_2A^+ \quad (7)$$

$$HA \xrightarrow{h\nu} B + \text{products} \quad (8)$$

$$B + HA \longrightarrow A^- \quad (9)$$

where B = a Bronsted base

We have recently carried out experiments using deuterated amino acids [5] to evaluate the possibility of molecular ion production by intermolecular proton transfer. Chain deuterated phenylalanine-d8 (1) and functional group deuterated phenylalanine-d3 (2) were used.

[Structures of chain-deuterated phenylalanine-d8 (1) with CD_2CD, NH_3^+, COO^- groups and functional group deuterated phenylalanine-d3 (2) with CH_2CH, ND_3^+, COO^- groups]

At threshold power densities, the major peak in the positive ion LMS of phenylalanine-d8 (1) was $(M + H)^+$; $(M + D)^+$ did not contribute more than 3% of the molecular ion current. Similarly, in phenylalanine-d3 (2) the peaks in the molecular ion region were primarily due to $(M + D)^+$; $(M + H)^+$ contributed no more than 7% in the molecular ion region. The negative ion spectra showed exclusively $(M - H)^-$ for phenylalanine-d8 and $(M - D)^-$ for phenylalanine-d3.

Another important experiment was to observe the laser mass spectra of deuterated amino acids at power densities several hundred times threshold values; peaks due to both H^+ and D^+ were observed in the spectra. Under these conditions the percentage of $(M + D)^+$ ions for phenylalanine-d8 in the molecular ion region increased to 8.3%, even though the measured D^+/H^+ ratio was 2:1. On the basis of statistical protonation, the anticipated $(M + D)^+/(M + H)^+$ ratio would be 2.0; that observed was 0.11. Similar results were observed for phenylalanine-d3 and valine-d3.

The above experiments lead one to conclude that random protonation reactions can account for no more than 7% of quasimolecular ion production in the amino acids. This result correlates with simple bond energy calculations, which indicate that the energy required for concerted proton transfer (5) is at least 5-7 eV lower than that required for random protonation (6-7). Thus, the above study provides reasonable support for the idea that quasimolecular ion formation in laser mass spectrometry can occur by pair production from simple proton transfer reactions.

4. Internal Salts

Quaternary ammoniohexanoates and their corresponding sulfonate analogs, sultaines, show group transfer reactions in both positive FD and in laser mass spectrometry [6,7]. These could occur by:

$$\underset{\overset{|}{CH_3}}{\overset{\overset{|}{CH_3}}{R-N^+}}-(CH_2)_nCOO^- \xrightarrow{h\nu} \underset{\overset{|}{CH_3}}{\overset{\overset{|}{CH_3}}{R-N^+}}-(CH_2)_nCOOCH_3 + \underset{\overset{|}{CH_3}}{\overset{\overset{|}{CH_3}}{R-N}}-(CH_2)_nCOO^- \qquad (10)$$

where $R = C_nH_{2nm}$ (n = 1-24)

or as an alternative:

$$\underset{\overset{|}{CH_3}}{\overset{\overset{|}{CH_3}}{R-N}}-(CH_2)_nCOO^- \xrightarrow{h\nu} CH_3^+ \ ; \ \underset{\overset{|}{CH_3}}{\overset{\overset{|}{CH_3}}{R-N}}-(CH_2)_nCOO^- + CH_3^+ \longrightarrow \underset{\overset{|}{CH_3}}{\overset{\overset{|}{CH_3}}{R-N^+}}-(CH_2)_nCOOCH_3 \qquad (11)$$

Transfer of the R group is also observed in laser mass spectrometry. Thus, the question arises whether transfer of a methyl group can occur by a process comparable to proton transfer in the amino acids. For the ammoniohexanoates studied (n = 1,8,12,24) the $(M + CH_3)^+$ peak corresponded to base peak for n = 1, 8, and 12 and was 50% of base for n = 24. The $(M - CH_3)^-$ peaks corresponded to approximately 70% of base for the four compounds. Thus, it is clear that methyl group transfer corresponds to an important process in laser mass spectrometry.

Addition of a small amount of p-toluene sulfonic acid (pTSA) to the internal salts causes the spectra to be dominated by $(M + H)^+$ ions; methyl group transfer reactions are very weak under these conditions, and alkyl transfer reactions are not observed. The process in the presence of pTSA is probably to desorb $(M + H)^+$ ions already present, as indicated by the drop in threshold energy for $(M + H)^+$ ions when pTSA is present. This competing equilibrium in the system prior to ionization is consistent with the methyl group transfer mechanism.

The threshold intensities of different ions from the C12 hexanoate vary as a function of laser power. The threshold for $(M + R)^+$ ions is lower than for $(M + CH_3)^+$. If $(M + CH_3)^+$ occurs by the two-step reaction of Eq. 11, one would expect ion production involving the weakest bond to occur at the lowest laser power. Methyl group transfer is observed at ~ 0.14 µJ; alkyl transfer at ~ 0.22 µJ. The order of bond energies is: $CH_3-N > RCH_2-N$, the exact opposite of what would be anticipated. Also, the peak intensities from methyl group transfer are greater than would be anticipated statistically by the number of CH_3 and R groups in the molecule, indicating a concerted reaction. Deuteration studies indicate that $(M + CH_3)^+$ ions come exclusively from methyl group transfer, as opposed to rearrangement of the hexanoate chain. At low laser powers neither $(M + H)^+$ nor $(M + CH_3)^+$ ions are observed. As one increases the power density in the system, first $(M + H)^+$ ions appear; ions corresponding to methyl transfer are observed at higher laser powers. Neither H^+ nor CH_3^+ ions were observed at any laser power.

It is proposed that $(M + H)^+$ and $(M + CH_3)^+$ ions arise from internal transfer reactions, consistent with the principles of concerted organic reactions. It is assumed that as more energy is put into the system (laser power-density is increased), more energetic reactions will occur. By using simple bond-energy calculations for H-atom transfer reactions, one would predict the following energy sequence for ion thresholds:

$$\underset{\overset{|}{CH_3}}{RN}=CH_2^+ < (CH_3)_2N=CH_2^+ < (M + H)^+ \qquad 0.17 \text{ eV} < 0.82 \text{ eV} < 2.07 \text{ eV}$$

Using the same considerations, proton production requires:

$$RH \rightarrow H^+ + e + R \qquad E = 18.4 \text{ eV} \tag{12}$$

Thus, because $(M + CH_3)^+$ ions appear in the same energy region as $(M + H)^+$ ions and production of CH_3^+ is comparable to H^+ production in Eq. 12, the above calculations clearly support the idea of concerted methyl group transfer reactions.

Again, although the data are even more indirect than for the amino acids, results for the zwitterions indicate that methyl group transfer in a single step is a viable mechanism for $(M + CH_3)^+$ and $(M - CH_3)^-$ production.

5. Phenylbenzoate

The laser mass spectra of phenylbenzoate show the following major ions at threshold:

positive: $C_6H_5^+$ (70%), $C_6H_5\text{-}\overset{+}{CO}$ (100%), and $(M - CO_2-H_2)^+$ (20%);

negative: $C_6H_5O^-$ (100%), $C_6H_5\text{-}COO^-$ (40%), and $(M - H)^-$ (25%).

Two reactions involving intramolecular ion-pair formation can account for production of the four major ions:

$$C_6H_5COOC_6H_5 \xrightarrow{h\nu} C_6H_3CO^+ + C_6H_5O^- \qquad E = 8.9 \text{ eV} \tag{13}$$

$$C_6H_5COOC_6H_5 \xrightarrow{h\nu} C_6H_5^+ + C_6H_5COO^- \qquad E = 9.2 \text{ eV} \tag{14}$$

We are currently involved in a study of phenylbenzoate to determine if charge separations such as those indicated above represent a feasible ionization mechanism for laser mass spectrometry.

Although the data show considerable scatter, the intensity ratio $[\phi O^-]/[\phi COO^-]$ is constant at 2.8 ± 1 as a function of laser power in the range of 1x-3x threshold. Similarly, a plot of $[C_6H_5CO^+]/[C_6H_5^+]$, as a function of laser power, gives an average value of 2.6 ± 1 in the same range, although there is a positive slope to the plot. The similarity of these numbers tends to support the idea that ion production is occurring according to Equations 13 and 14.

The energy required for Eqs. 13 and 14 is 8.3 and 9.2 eV, respectively. The ionization potential of phenylbenzoate is 8.98 eV. Thus, both reactions require approximately the same energy as molecular ion formation. Therefore, one must consider two possible reaction channels for producing these ions. Two photons (265 nm; 4.68 eV) are barely sufficient to bring about ionization of phenylbenzoate (1.92 photons required). A number of excited electronic states of phenylbenzoate are accessible to a single photon, thus a three-photon process is required by either channel. Thus, one is in reality considering either photodissociation of the molecular ion, or photodissociation of an excited electronic state.

Photodissociation of the molecular ion of phenylbenzoate can occur in several ways:

$$[C_6H_5COOC_6H_5]^+ \xrightarrow{h\nu} C_6H_5CO^+ \qquad E = 1.3 \text{ eV} \tag{15}$$

$$[C_6H_5COOC_6H_5]^+ \xrightarrow{h\nu} C_6H_5^+ \qquad E = 3.7 \text{ eV} \tag{16}$$

$$C_6H_5CO^+ \xrightarrow{h\nu} C_6H_5^+ \qquad E = 2.7 \text{ eV} \tag{17}$$

Consideration of ion production from $M^{+\cdot}$ introduces three other possible channels for the formation of the ions observed. Contrary to electron impact MS, the molecular ion of phenylbenzoate is not observed in LMS even at threshold. However, one cannot be certain that molecular ions are not formed. Thus, it is important to consider the relative rate constants for Equations 13-17 to determine which represent the more likely routes.

We have calculated the RRKM rate constants as a function of energy for reactions 15 and 16. At energy levels below 7 eV, the RRKM rate constant for Eq. 15 is approximately 10^5 that of Eq. 16. Although such calculations are only approximate, this large difference indicates that Eq. 16 does not contribute significantly to the production of $C_6H_5^+$ ions in phenylbenzoate. Calculations also indicate that the rate constant for Eq. 17 should increase with increasing energy predicting a decrease in the $[C_6H_5CO^+]/[C_6H_5^+]$ ratio as a function of input energy. This effect is observed experimentally. Thus, one can best interpret ion formation in phenylbenzoate by internal ion-pair production according to Eqs. 13 and 14. We are currently in the process of using deuterated compounds to obtain a more exact measure of the extent of reaction due to each channel.

6. Summary

In summary, it is proposed that direct production of ion pairs is a viable mechanism for laser mass spectrometry. From the evidence accumulated to date, it appears that this can occur both by group transfers (H^+, CH_3^+) and fragmentation of a molecular chain to produce an ion pair. It is quite likely that excited electronic states are involved in these processes, and that within any given system more than one channel for ion production is available.

7. Acknowledgment

The research on which this report is based was supported by the National Science Foundation and the Office of Naval Research.

References

1. D. M. Hercules, "Solid State Mass Spectrometry Using a Laser Microprobe" in K. J. Voorhees, Ed., Analytical Pyrolysis Techniques and Applications, pp. 1-41, Butterworth, London, 1984.

2. D. M. Hercules, Pure and Appl. Chem., 55, 1869 (1983).

3. N. Omichi, J. Silverstein, R. D. Levine, J. Phys. Chem., 85, 3364, 1981.

4. R. S. Berry, T. Cenoch, N. Coplan, J. J. Ewing, J. Chem. Phys., 49, 127, 1968.

5. C. D. Parker and D. M. Hercules, Anal. Chem., 57, 698, 1985.

6. C. D. Parker and D. M. Hercules, Anal. Chem., in press.

7. T. Keough, A. J. DeStefano, R. A. Sanders, Org. Mass Spectom., 15, 351, 1980.

8. K. Balasanmugam, D. M. Hercules, Anal. Chim. Acta, 166, 1, 1984.

9. H. Nguyen, D. M. Hercules, unpublished studies, University of Pittsburgh, 1985.

The Influence of the Substrate on Ultraviolet Laser Desorption Mass Spectrometry of Biomolecules

F. Hillenkamp[1], D. Holtkamp[2], M. Karas[1], and P. Klüsener[2]

[1]Institut für Biophysik, Universität Frankfurt,
Theodor-Stern-Kai 7, D-6000 Frankfurt/M. 70, F.R.G.
[2]Physikalisches Institut, Universität Münster, Domagkstr. 75,
D-4400 Münster, F.R.G.

1. Introduction

Several investigators have reported that for Laser Desorption Mass Spectrometry (LDMS) with pulsed infrared lasers and foci of about 1 mm the absorption and/or heat conduction properties of the substrate have a strong influence on the spectra obtained [1, 2, 3]. For these types of LDMS, resonant absorption of the laser radiation by the sample has also been shown to effect the types and intensity of desorbed ions [4, 5]. The influence of resonant absorption and of the substrate properties either singly or as combined effects are as yet unknown, or discussed controversially for UV-LDMS [6, 7, 8].

2. Experimental

In order to keep sample preparation controllable and reproducible, samples of the amino acid L-tryptophane (trp) of varying thickness were produced by a molecular beam technique, evaporating the sample material onto substrates of etched silver or quartz. For comparison, samples were also prepared by vacuum drying 1 µl of a 10^{-2} mol/l solution onto silver substrates. The beam source was a resistance-heated stainless steel oven with a cylindrical opening of 13 mm length and 2 mm diameter [9]. Its temperature was monitored by a thermocouple and controlled to $\pm$ 2 K throughout each run. A quartz crystal microbalance, mounted slightly off axis, next to the substrate, was used to measure the deposited areal density (m/A). The overall accuracy of this measurement is estimated to $\pm$ 20%. Samples with areal densities in the range of $2 \cdot 10^{-7}$ g/cm^2 (nominal monolayer) to $2 \cdot 10^{-4}$ g/cm^2 were produced. The Ag-substrates (Degussa finesilver) were etched in a 20% HNO_3 solution for 0.5-1 min. at 50 $^\circ$C under sonication. Quartz substrates of optical quality were cleaned in H_2O, methanol and freon, also under sonication. The tryptophane was commercial powder material from E. Merck A.G.

Absorption spectra of the samples on quartz substrates were taken in the wavelength range of 200-500 nm with a Cary 15 spectrophotometer, with sample and reference beam apertured down to 3 mm. All mass spectra were taken with a LAMMA 1000 instrument. The laser pulse width was app. 10 ns, the frequency quadrupled line of a Nd-YAG laser at 265 nm wavelength was used in most cases. In a series of control experiments the frequency tripled line at 355 nm was used. The laser beam was incident on the sample surface at an angle of 45° to the surface normal, the ions were extracted along the normal. Samples on quartz substrates were covered with a fine wire mesh, to prevent charging of the specimen. A more detailed description of the LAMMA 1000 instrument can be found elsewhere [10].

3. Results

Fig. 1 shows three absorption spectra of tryptophane evaporated onto quartz substrates at three different oven temperatures of 165 °C, 190 °C and 220 °C, as well as a spectrum of a $8.3 \cdot 10^{-5}$ mol/l tryptophane aqueous solution for comparison. All spectra exhibit the typical absorption band of the indole S_o-S_1 transition around 280 nm. The absorption peak of the evaporated samples is redshifted by about 5 nm (385 nm) relative to that of the solution. The fine structure of the band is conserved and, within the measurement uncertainty, the measured peak absorbance of the evaporated samples has the expected absolute value for the measured areal density, assuming a molar extinction coefficient of $\varepsilon_n = 5.5 \cdot 10^3$ $l \cdot mol^{-1} \cdot cm^{-1}$ as reported in the literature for trp in solution. For oven temperatures above 165 °C the spectra show considerable absorption throughout the near UV and into the Visible. This additional absorption does not follow a $1/\lambda^4$ relationship, so it is not caused by scattering of light in the not perfectly homogeneous layers.

At irradiances near threshold for the desorption of sample specific ions, spectra of all three different preparations are identical in the mass range of M_r/Z= 205 and

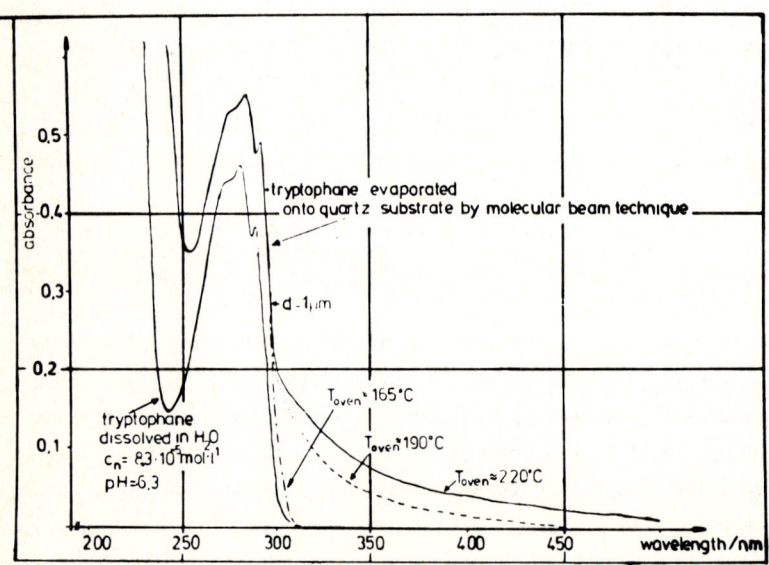

Fig. 1 Absorption spectra of tryptophane in solution and tryptophane, evaporated onto quartz substrates at different temperatures.

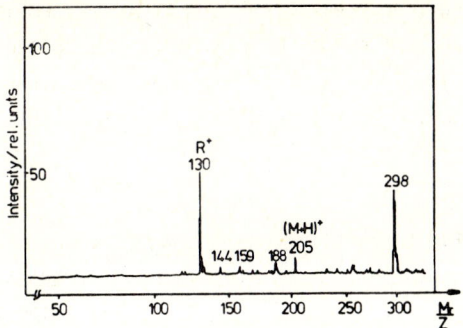
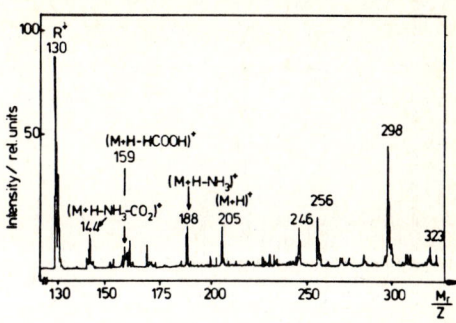

Fig. 2 Positive ion spectrum of tryptophane, evaporated onto an Ag-substrate (t_{oven} = 190 °C; d= 1.0 μm) taken at a near threshold irradiance (λ= 265 nm)

Fig. 3 Positive ion spectrum of tryptophane evaporated onto a quartz substrate (t_{oven} = 190 °C; d= 1 μm) taken at a near threshold irradiance (λ= 265 nm)

less. Figs. 2 and 3 show the positive ion spectra of two samples evaporated onto Ag- and quartz substrates. Base peak is the aromatic residue R^+ at M_r/Z= 130. Besides the protonated parent molecule $(M+H)^+$ at M_r/Z= 205 some typical fragments and decomposition products $((M+H-NH_3)^+$, 188; $(M+H-HCOOH)^+$, 159; $(M+H-NH_3-CO_2)^+$, 144) occur. The threshold irradiance of $2\cdot 10^7$ W/cm^2 was the same for all three preparations and below that for the generation of Ag^+-ions out of the uncovered Ag-substrates by a factor of 5. Because of this low threshold it is not surprising to not find any Ag-ions in the spectrum. However, even at an irradiance a factor 8 above threshold, sufficient to generate Ag^+-ions, Ag-signals are lacking in the spectra. Ag-peaks occur only in the 3. to 11. spectrum, taken from the same sample area, depending on the sample thickness (0.01 μm-1 μm) (Fig. 4). Only for samples of m/A= $2\cdot 10^{-7}$ g/cm^2, nominally a monolayer, did Ag-signals occur in the first shot. In these cases the threshold irradiance for trp-signals also matched that for Ag^+-ions. All spectra taken from these thin layers contain the residue R^+-peak, some of them even all the other trp-signals shown in Figs. 2 and 3 from thicker layers.

The spectra of Figs. 2, 3 and 4 all were obtained from samples that had been evaporated at a temperature of 190 °C. All show a strong peak at M_r/Z= 298 and some weaker ones at 310 and 323 M_r/Z. These peaks never occurred in any of the spectra from samples evaporated at 165 °C. These peaks are very stable even at irradiances at which no trp-signals can be identified any more.

For comparison, spectra of trp-samples, evaporated at t= 165 °C onto Ag-substrates were also obtained at a laser wavelength of 353 nm (Fig. 5). Even at threshold irradiance, the spectra always show strong signals of silver, besides the usual trp-signals. In contrast to the spectra obtained at 265 nm, the ones taken at 353 nm show a

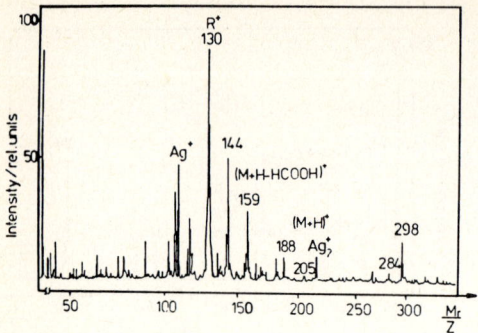
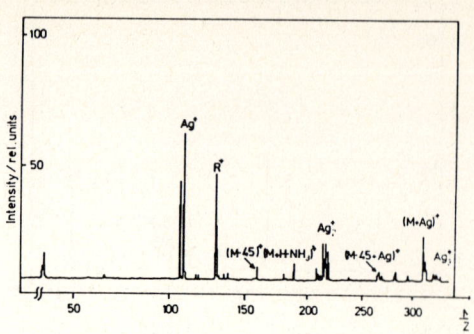

Fig. 4 Positive ion spectrum of tryptophane evaporated onto an Ag-substrate (t_{oven} = 190 °C; d = 0.1 μm) taken at a near threshold irradiance (λ = 265 nm)

Fig. 5 Positive ion spectrum of tryptophane evaporated onto an Ag-substrate (t_{oven} = 165 °C; d = 0.014 μm) taken with 353 nm laser radiation at a near threshold irradiance

considerable abundance of Ag-cationized trp and fragments thereof. The threshold irradiance for trp at this wavelength is about a factor 50 higher than at 265 nm as expected [8].

4. Discussion

The optical absorption spectra prove that the transition from solution to the solid state introduces only very small changes in the absorption bands. Though this finding is not unexpected, considering the solid state structure of the molecule, it´s experimental verification was an essential prerequisite for linear energy transfer calculations, as are presented e.g. by Karas et al. [8].

At the wavelength of 265 nm, substrate absorption does not contribute to the desorption of trp-ions. This is evidenced by equal threshold irradiances and identical spectra for absorbing and nonabsorbing substrates. The strong absorption of trp at 265 nm explains it´s relatively low threshold irradiance, because calculations [8] show that at this irradiance each trp-molecule in the top sample layer absorbs as much as about 20 eV, certainly enough for desorption. Rather surprising is, however, the fact that even for sample layers as thin as 0.01 μm, which should transmit 92% of the incident energy down to the substrate, no Ag-signals were ever seen in the first shot, even though in these experiments the irradiance had been raised above the threshold for Ag^+-ions. It is suggested, that the high laser irradiance induces a strong nonlinear absorption, which shields the underlying substrate. The complete lack of Ag-cationization in spectra of successive shots which show trp- and Ag-signals seems to indicate that the trp-ions and Ag^+-ions are generated at different locations within the laser spot of app. 3 μm in diameter, the former most probably at the rim, the latter in the center.

The results obtained at 353 nm support the suggested model. Because of lack of absorption, enough energy penetrates down to the substrate, even through layers of 0.1 μm. Ag^+-ions are therefore very abundant in every spectrum taken. Even though the threshold irradiance at this wavelength is about a factor of 50 above that at 265 nm, nonlinear absorption does not yet set in. The reason for this is the strong wavelength dependence of the field strength required for multiphoton processes and the well-documented decrease of threshold irradiance for the onset of such effects if there is natural absorption. The abundant cationization seen in the spectra taken at 353 nm suggests that trp- and Ag^+-ions are either generated in the same locations, or mix well above the sample surface.

The mass peaks around M_r/Z= 300 for trp-samples, evaporated at temperatures of 190 °C together with the absorption, extending into the near UV and Visible suggests the formation of condensed aromates most probably from the indole residue through chemical reactions with the oven walls.

References

1. F. Hillenkamp: "Laser-Induced Ion Formation from Organic Solids" (Review), in Ion Formation from Organic Solids, Springer Series in Chemical Physics 25, A. Benninghoven, ed. Springer Verlag 1983

2. G.J.Q. van der Peyl, J. Haverkamp and P.G. Kistemaker: Int. J. Mass Spectrom. Ion Phys. 42 (1982) 125-140.

3. B. Schäfer and P. Hess: Int. J. Mass Spectrom. Ion Proc. 47 (1984) 47-50.

4. R. Stoll and F.W. Röllgen: Z. Naturforsch. 37a (1982) 9.

5. M. Mashni and P. Hess: Chem. Phys. Lett. 77 (1981) 541.

6. D.M. Hercules, R.J. Day, K. Balasanmugam, T.A. Dang and C.D. Li: Anal. Chem. 54 (1982) 280A-290A.

7. V.S. Antonov, V.S. Letokhov and A.N. Shibanov: Appl. Phys. B28 (1982) 245.

8. M. Karas, D. Bachmann and F. Hillenkamp: Anal. Chem., in print.

9. D. Holtkamp, W. Lange, M. Jirikowsky and A. Benninghoven: App. Surf. Science 17 (1984) 296-308.

10. P. Feigl, B. Schueler and F. Hillenkamp: Int. J. Mass Spectrom. Ion Phys. 47 (1983) 15-18.

On Different Desorption Modes in LDMS

B. Lindner and U. Seydel

Forschungsinstitut Borstel, Institut für Exp. Biologie und Medizin,
D-2061 Borstel, F. R. G.

1. INTRODUCTION

In laser desorption mass spectrometry (LDMS) two different configurations concerning the arrangement of the laser and of the mass analysing system with respect to the sample are commonly used. In reflection- type instrumentation, irradiation and extraction of the desorbed ions are performed on the same side of the sample, in transmission-type instruments on opposite sides. This paper focuses exclusively on the latter type, particularly on the laser microprobe mass analyser LAMMA 500 (Leybold-Heraeus, Köln, F.R.G.).

In a recent paper [1] we have described two extremely different desorption modes characterized by the degree of thermal stress exerted to the desorbed molecules which can be controlled via sample thickness and laser irradiance and which are obviously subject to different mechanisms.

From the observed differences in the fragmentation patterns of thermolabile substances it seemed to be reasonable to expect differences in ion kinetic energies for the different modes and ion species. To get this information we adopted experimental techniques described by MAUNEY et al. [2], who utilized the ion reflector of the TOF as an energy cut-off filter and by WURSTER and WIESER [3] who applied a field-free drift path to determine kinetic energies from the shift in the arrival time of the ions at the detector.

2. Experimental

The parameter settings of the LAMMA 500 – for a detailed description of the instrument see e.g. [4] – for the present investigations were as follows: wavelength 265nm, pulse duration 10ns, laser power density 10^9 to 10^{11} Wcm^{-2}.

Samples were prepared from mixtures of aqueous solutions of sugars (stachyose, α-cyclodextrin) and alkali salts (NaI, CsI) at a molar ratio of appr. 5:1. Different amounts of these solutions were brought onto Formvar coated copper grids and dried. This way layers of different thicknesses could be achieved guaranteeing, to a far extent, homogeneity in thickness and consistency over the distance of several meshes.

In the reflectron-type TOF analyser, the retarding potential is equal to the sum of the accelerating and reflector voltages. At each setting of the reflector voltage, only those ions with initial kinetic energies in direction toward the TOF smaller than the value of the reflector voltage are reflected and detected. To get information on the kinetic energies of the different ion species under various experimental conditions, the reflector voltage was varied between −150V and +200V. At each setting 30 spectra from the same sample were registered and averaged.

In a second experiment, a field-free drift path with a length of 2mm was introduced directly adjacent to the back surface of the sample. As a consequence of this, the position of the mass peak will be shifted with respect to the arrival time at the detector. From this shift initial kinetic energies can be calculated.

3. Results

The experimental parameters commonly used with the LAMMA 500 are thin sample layers and low laser irradiance just above the threshold of sample perforation, typically 10^7 Wcm^{-2}.

In contrast to this condition we favour high laser power densities (10^{11} Wcm^{-2}) and thick sample layers (20μm) which are normally not perforated. From some substances of high molecular weight, only under these conditions could spectra be produced. The relatively simple spectra with abundant quasimolecular peaks and low degree of fragmentation even from thermolabile compounds point to a very gentle desorption/ionization process which was explained by a shock-wave model [1].

Differences in the spectra obtained under these two sets of parameters are shown in fig. 1 for stachyose. It could be demonstrated futhermore, that intermediate conditions lead to spectra combining both characteristics.

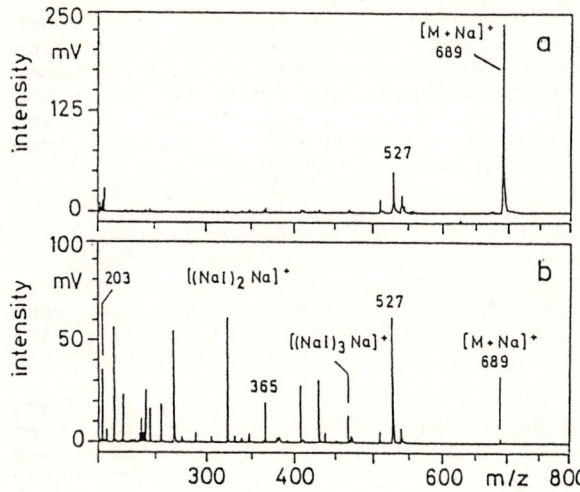

Fig.1: Averaged positive ion LD-mass spectrum of a mixture of stachyose and NaI obtained for:
a) d=20μm, p=10^{11} Wcm^{-2}
and b) d=1μm, p=10^8 Wcm^{-2}

Figure 2 gives the ion intensities in spectra of stachyose and α-cyclodextrin mixed with CsI in dependence on the reflector voltage obtained from a sample of medium thickness (10μm).

The different response of the various ion species on changes in the reflector voltage is obvious. The intensity ratio of a quasimolecular ion ([stachyose+Cs]$^+$) and a cluster ion ([Cs(CsI)$_2$]$^+$

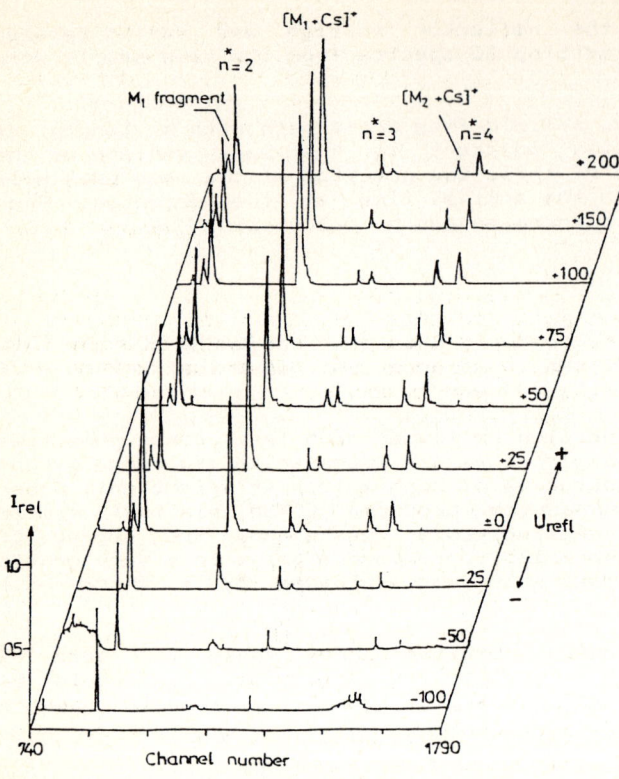

Fig.2: Positive ion LD-mass spectra of a mixture of stachyose (M1), α-cyclodextrin (M2) and CsI in dependence on the reflector voltage obtained from a medium thick sample layer (10μm); * :cluster ions of type $[Cs(CsI)_n]^+$

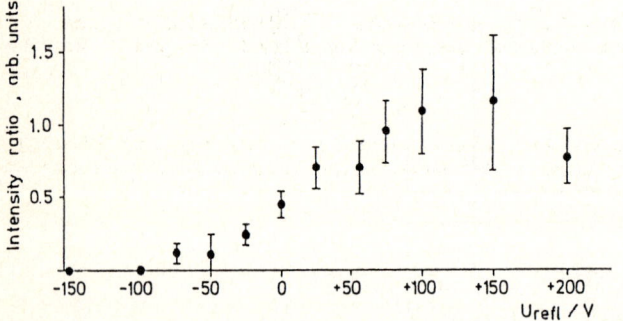

Fig.3: Ratio of the intensities of the quasimolecular peak of stachyose ($[M+Cs]^+$) and the cluster ion $[Cs(CsI)_2]^+$ vs. reflector voltage (calculated from fig.2)

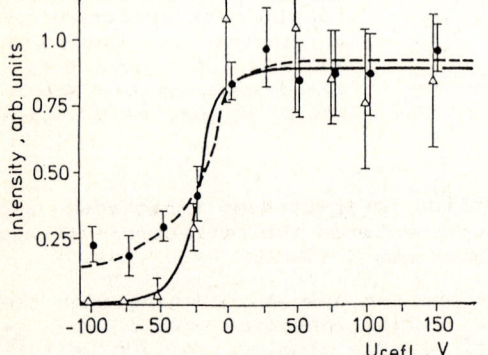

Fig.4: Intensity of the quasimolecular ion of stachyose vs. reflector voltage for thick (--) and thin (—) sample layers

vs. reflector voltage is plotted in fig. 3 demonstrating differences in the behaviour of these particular ions. However, all cluster ions of the type $[Cs(CsI)_n]^+$ on the one hand and all molecular and fragment ions on the other hand each show the same dependence.

The variations in the intensity of the quasimolecular peak of stachyose with the reflector voltage is depicted in fig. 4 for the two desorption modes of fig. 1 illustrating the more rapid intensity decrease in the case of a thin sample layer and moderate laser irradiance.

Figure 5 shows an example of the influence of the field-free drift path on the arrival time of the quasimolecular ion of stachyose for a thin sample irradiated at . $2*10^{10}$ Wcm^{-2}. The broadening of the shifted peak is due to the energy spread of the ions and masks the isotope distribution. The ion kinetic energies are calculated from the time shifts between the peak maxima.

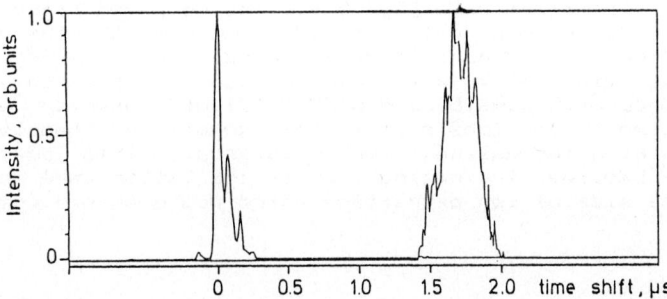

Fig.5: Shift in the arrival time of the quasimolecular ion of stachyose between the normal (left) and the modified (right) sample stage

This way, the influence of the laser irradiance on the ion kinetic energy for thick and thin sample layers was measured, as illustrated in fig. 6.

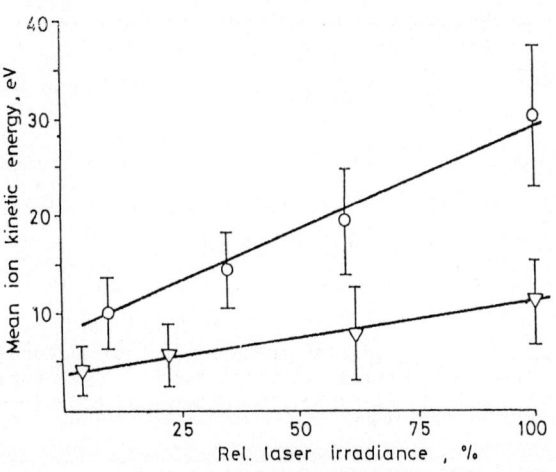

Fig.6: Mean kinetic energy of the quasimolecular ion of stachyose.vs. laser irradiance for thick (o) and thin (▽) sample layers (100% = 10^{11} Wcm^{-2})

4. Discussion

It can be statet that, in principle, both applied techniques show differences in the ion kinetics between the two above described desorption/ionization modes. However, there arise problems concerning the interpretation of the results from the energy cut-off measurements. The observed energy deficits, i.e. the ions arriving at the reflector have kinetic energies less than the acceleration potential leading to the registration of ion intensities even at negative reflector voltages, can not definitely be explained. Possible explanations could be, for instance, that the ions are generated not directly on the sample surface and thus do not 'see' the total acceleration potential, or that charge interactions between ions take place (for a futher discussion see also [2]). The clarification of these questions must remain subject to futher investigations.

The results obtained with the field-free drift path seem to be more readily understandable. The measured ion kinetic energies of the relatively large and complex molecules and clusters show an overall agreement with literature data [2,3,5,6]. The observed higher slope in the ion energy dependence on the irradiance for the thick sample may be explained by a more direct translational energy transfer to the ions via shock wave than in case of the thin samples. In accordance with WURSTER and WIESER [3] we observed a shift in the arrival time of the ions even in the unmodified sample stage for thin samples with increasing laser irradiance. This does not occur with thick samples, indicating that in the latter case no plasma is formed on the side of ion extraction which would shield the acceleration field.

5. References

1. B.Lindner, U.Seydel: Anal.Chem. 57, 895 (1985)
2. T.Mauney, F.Adams: Int. J. Mass Spectrom. Ion Phys. 59, 103 (1984)
3. R.Wuster, P.Wieser: Proc. of the 2nd LAMMA-Workshop (1983), p.21
4. H.J.Heinen, S.Meier, H.Vogt, R.Wechsung: Adv.Mass Spectrom. 8, 942 (1980)
5. R.Nitsche, R.Kaufmann, F.Hillenkamp, E. Unsöld, H.Vogt, R.Wechsung: Isr. J. Chem. 17, 181 (1978)
6. E.D.Hardin, M.L.Vestal: Anal. Chem. 53, 1492 (1981)

Part V

Other Ion Formation Processes

"Spontaneous" Desorption of Negative Ions from Organic Solids and Films of Ice at Low Temperature

S. Della-Negra[1], C. Deprun[1], Y. le Beyec[1], J. Benit[2], J.P. Bibring[2], and F. Rocard[2]

[1] Institut de Physique Nucléaire, B.P. N° 1, F-91406 Orsay Cedex, France
[2] Laboratoire René Bernas, B.P. N° 1, F-91406 Orsay Cedex, France

Introduction

It has been shown (1) that under the effect of a constant voltage applied between a flat metallic surface and a high transmission grid (90 %), electrons and negative ions are emitted simultaneously. Time-of-flight mass measurements can be made by taking electrons as start signals. Mass spectra of organic samples have been compared with those obtained by PDMS. The method - named by us "Spontaneous" Desorption Mass Spectrometry - has also been used to study the emission from a cooled ice target and its dependence with temperature (from 77 to 200 K). Experimental investigations on the mechanisms are presented.

Experimental

At one end of a time-of-flight tube a set of channel plate detectors is mounted. The signals from a constant fraction discriminator is fed in the start and stop inputs of a multistop TDC. The start and stop signals are thus the same, the "stop" being delayed by a few hundred nanoseconds. The data acquisition system is similar to that used for usual time-of-flight mass spectrometry. A fixed difference of potential (-5 to -12 KV) is applied on a distance of $\sim$ 4 mm between the sample and the grid at ground. For organic analysis, an aluminized mylar foil is coated with a few micrograms of compounds and the measurements are made with the "DEPIL" time-of-flight spectrometer (2). In the case of ice samples at liquid N_2 temperature we have used another time-of-flight system (3). Figure 1 shows the principle of the apparatus. Films of ice have been made under vacuum through a well-defined procedure (4).

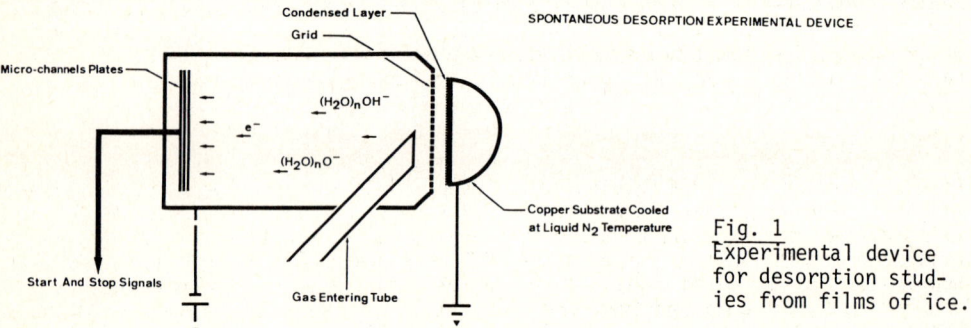

Fig. 1 Experimental device for desorption studies from films of ice.

Results

Organic solids

Several mass spectra of different organic compounds have been made and compared with spectra obtained by fission fragment ionization desorption (leu and met enkephaline, chloranphenicol, arginine... phosphazene). The spectra look very similar.

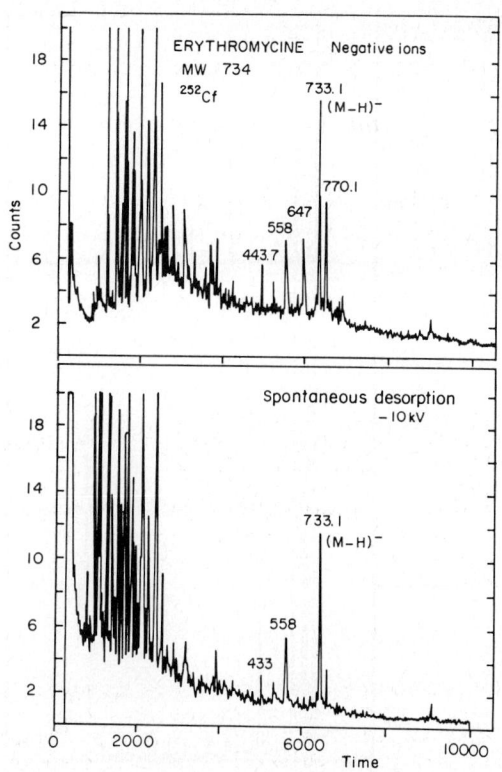

Fig. 2
Comparison of mass spectra obtained with fission fragment induced desorption and spontaneous desorption for an organic molecule.

More time is needed to record a spectrum by this technique because all electrons are not time correlated with negative ions. Also, large involatile molecules have not been observed yet. Figure 2 shows an example of spectra obtained with erythomycine. The molecular ion peak $(M-H)^-$ at m/z = 733 is clearly seen. In this technique the time-to-mass calibration is made with electrons (mass zero) and mass 25^- (C_2H^-) or H^- which is almost always present in the spectra. In the low mass region, 12^- and 13^- are also observed in very small quantity. It must be noted that the presence of peaks at m/z 16^- and 17^- are due to a shift in the mass spectrum which occurs when the time measurement is triggered by H^- instead of e^-. In this case the peaks at 25 and 26 appear at the position of 16 and 17.

Films of ice

Figure 3 shows a spectrum obtained with ice at 77 K. In this experiment the time-of-flight distance L is only 13 cm and the relatively good mass resolution $M/\Delta M \sim 300$ (with respect to the small L value) indicates that the negative cluster ions are emitted from the surface. Masses above 1000 have been observed. During a very slow heating of the condensed layer ($\sim$ 100 minutes), mass spectra were measured every 3 minutes, in the multispectrum mode, and stored on disks. Temperature (77 K to 200 K) and vacuum pressure were recorded permanently. It has been found that the mass spectra were strongly dependent on the temperature and vacuum. The nature of the surface, the structure of the solid and the binding forces are modified and these changes are reflected by the mass spectra. For example, the cluster structure disappears in the mass spectrum after 180 K. Above 160 K, there is a slow and continuous increase of the pressure, which reaches a maximum at $\sim$ 190 K. The presence of peaks around m/z = 115^- is related to the variation of temperature and pressure, and the yield of these ions which follows the same trend increases very much when water sublimates completely. The temperature dependence of O^-, OH^- and $(H_2O)(OH))^-$

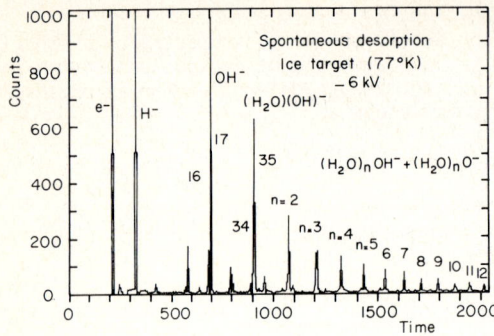

Fig. 3
Spontaneous desorption mass spectrum of a film of ice at 77 K. Only cluster ions with n < 12 are presented.

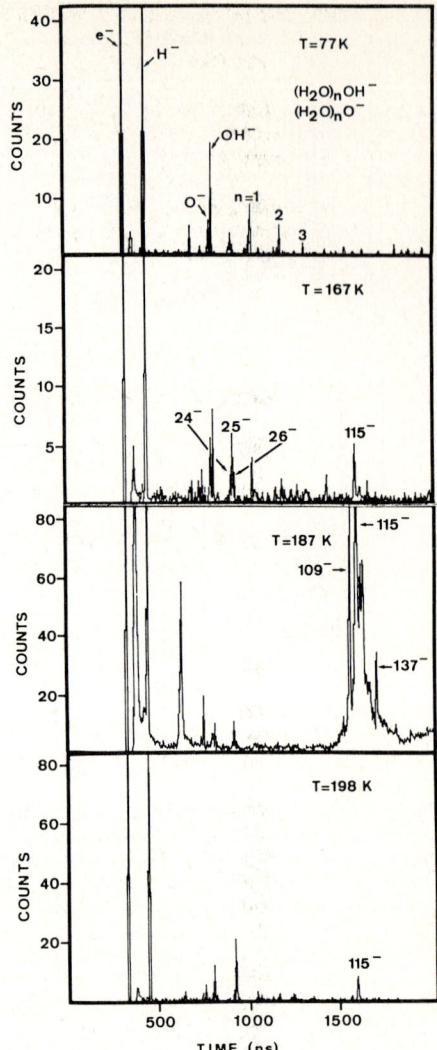

Fig. 4
Spectra measured at different ice temperatures. Masses around 115⁻ are characteristic of the sublimation process.

is very different. For these ions a strong increase of the yield is observed between 175 K and 185 K. At 200 K the peaks at m/z = 115⁻ are still slightly visible in the mass spectrum and above 200 K mainly peaks from contaminants (a turbo molecular pump was used) are observed.

Figure 4 shows four spectra corresponding to different mean temperatures. Masses around m/z 115⁻ have not been identified yet, but they are characteristic of the sublimation process. It is clear from this simple experiment that surface modifications which cannot be due to any ionization-desorption method are easily detected. It is also an indication that "spontaneous" desorption is largely enhanced when molecular surface bondings are weak.

Experimental investigations of mechanisms

Figure 5 shows spectra obtained with a simple experimental set-up to investigate possible explanations for the effects of simultaneous emission of electrons and

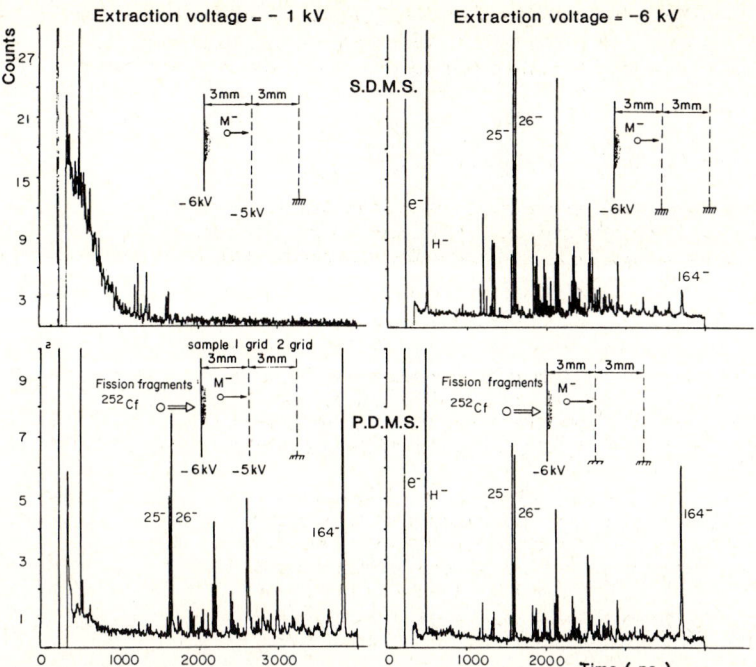

Fig. 5
Spontaneous and ^{252}Cf mass spectra of phenylalanine measured under different experimental conditions (see text).

negative ions. The right part of this figure shows a direct comparison of fission fragment-induced desorption and spontaneous desorption. Two grids 3 mm apart are at ground, the first grid being at a distance of 3 mm from the sample surface maintained at -6 KV. Molecular ion peaks (M-H)⁻ = 164 (phenylalanine) are seen in both spectra. Time-of-flight spectra are measured under exactly the same experimental conditions. The ^{252}Cf source can be placed behind the sample for the ^{252}Cf measurements and removed afterwards. The number of start obtained by integration of the e⁻ peak is used for normalization of the counting rate. Therefore one can see that the yield with ^{252}Cf is about 8 times larger.

On the left part of the figure, spectra have been recorded with only 1 KV of extraction voltage : sample at -6 KV, first grid at -5 KV and second grid at ground. The molecular peak at 164 is observed with ^{252}Cf but no molecular ions are desorbed with a voltage of 1 KV between the sample and the first grid.

If positive ions were present in the vacuum chamber or extracted from the second grid, they could be accelerated by the field between the two grids and hit the sample with an energy of 6 keV, generating the emission of electron and negative ions. Another explanation could be that an electron emitted from the sample hit the grid generating positive ions, which then fall on the sample after being accelerated. In the case of two grids, electrons can be extracted from the first grid as well, and hit the second one. Since no molecular ions are observed, it is difficult to believe in these hypotheses. Furthermore, the electric transmission of the grids is rather large (> 80 % for an optical transmission of 90 %) and it is known that the secondary ion yield for keV ions lies between 10^{-3} and 10^{-4}. It would therefore be difficult to understand the spontaneous desorption ion yield (with respect to the number of electrons) which can reach 10^{-2}.

We believe that preformed ions exist on the surface or are formed via charging on the surface layer, and that emission is probably taking place from microscopic protuberances which always exist on a surface.

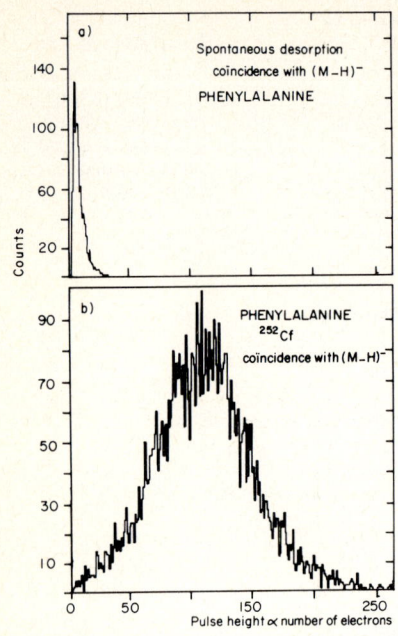

Fig. 6
Number of electrons associated with the simultaneous desorption of the molecular ion (M-H)⁻ of phenylalanine in the two cases :
- fission fragment-induced desorption (30-40 e⁻)
- spontaneous desorption (1-2 e⁻).

Since electrons and ions are desorbed simultaneously, it is possible to estimate the number of electrons leaving the surface with a molecular ion. A simple electronic set up allows to analyse heights from the channel plate detectors in coincidence with a given mass (5). It has been shown that the amplitudes are proportionnal to the number of primary electrons (5,6). When a fission fragment passes through the sample, it generates the emission of ions and many electrons(up to 40 e⁻ per fission fragment). With the same sample of phenylalanine we have measured the amplitude response of the channel plates for electron impact when a molecular ion (M-H)⁻ = 164 is desorbed and detected later on. For comparison we show in Fig. 6 the electron number spectrum in the case of spontaneous desorption and in the case of fission fragment desorption. The position and the shape of the electron S.D spectrum indicate that mainly one or two electrons are emitted when (M-H)⁻ is emitted. If electrons can play a role in fission fragment (or fast heavy ion) desorption mechanisms, it is obvious that it is not the only mechanism involved. It seems to us that there is a mixture of several mechanisms, and that negative organic or cluster ions which can be seen through the S.D method are also easily observed with fast heavy ions. Cristalline or amorphous structures of the surface are certainly of importance, and understanding the processes requires further investigation.

References

(1) S. Della-Negra, P. Häkansson, Y. Le Beyec, Nucl. Instr. and Meth. 89 (1985) 83
(2) S. Della-Negra, C. Deprun, Y. Le Beyec, IPNO-DRE-85-22.
(3) J. Benit, J.P. Bibring, S. Della-Negra, Y. Le Beyec, F. Rocard, in press Nucl. Instr. and Meth. in Physic Research.
(4) J.P. Bibring and F. Rocard, Adv. Space Res. 4 (1984) 102
(5) S. Della-Negra, D. Jacquet, I. Lorthiois, Y. Le Beyec, O. Becker and K. Wien, Int. J. of Mass Spectr. and Ion Processes 53 (1983) 215.
(6) O. Becker, S. Della-Negra, Y. Le Beyec, K. Wien, IPNO-DRE-85-21.

Electric Pulse-Induced Desorption Compared to Other Techniques - Mechanism, Mass Spectra, and Applications

F.J. Mayer, F.R. Krueger, and J. Kissel

Max-Planck-Institut für Kernphysik, Abt. Kosmophysik (Dust Group),
Postfach 10 39 80, D-6900 Heidelberg, F. R. G.

1. Introduction

It is well known that even very large molecules can be desorbed from a solid, when a transient but strong electric perturbation affects its surface, regardless of the detailed properties of the source of excitation. This can be done either by exciting the electronic system of the solid directly, e.g.,
- with very fast heavy ions in the inelastic energy loss regime (HIID)
- with non-resonant UV-laser beams (LID)

or, by exciting the lattice structure, thus altering the electric environment of the molecules at the surface rapidly, e.g.,
- with primary ions in the elastic energy loss regime (SIMS)
- with far-IR lasers resonant to lattice modes.

A third experimental method of exciting the lattice for molecular desorption was found recently [1]: the dust particle impact method (PaID). However, with these lattice excitation methods, sputtering, i.e., direct momentum transfer to atoms and small molecules, is a common competing process.

In contrast, the scope of this paper is to deal with excitation functions and other quantitative features of the electronic-type excitation modes. To do this, a third electronic excitation method, the "Electric Pulse Induced Desorption" (EPID), has been introduced [2], which technique is very simple and thus useful for a number of fundamental investigations, although it suffers from several drawbacks concerning analytical applications.

Restricting oneself to electronic excitations, one gets rid of ion formation processes such as direct sputtering, competing with the very process under investigation, namely the "diabatic" desorption, i.e., the prompt emission of large molecules and fragments from polar substances. Furthermore, using time-of-flight (TOF) spectrometers, one can easily discriminate another competing process, i.e., the simple thermal evaporation of (especially alkali-) ions, which lasts micro- to milli-seconds, as long as the excited surface is hot enough. Finally, restricting ourselves to only low heavy-ion fluxes (in HIID), low laser fluences (in LID), and low currents (in EPID), respectively, the measurements are not disturbed by another competing process, the formation of a collision plasma by the secondary particles.

The process of interest is diabatic desorption - or "diabatic sputtering", as it was called by KELLY and ROTHENBERG [3] -, as introduced by one of us (F.R.K.) [4,5] several years ago. This process, being explained in terms of non-adiabatic quantum (time-dependent) perturbation theory, is made responsible for the emission of even large molecular ions from the uppermost layers of a bulk, the forecast properties of which being widely confirmed, e.g., by recent measurements by KELLY and DREYFUS [3,6].

2. The Mechanism of Diabatic Desorption

As shown in Fig. 1, the electronic interaction of a free molecule with a surface can be represented by splitting the potential state of the whole molecule into two

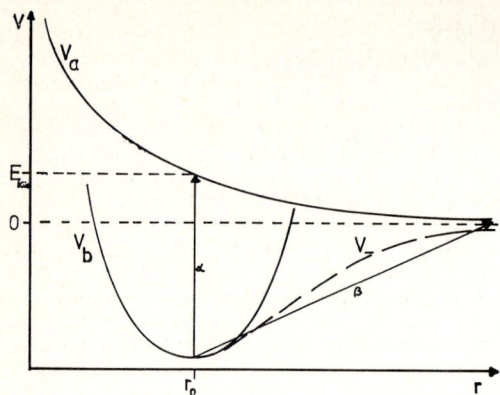

Fig. 1 Attractive and repulsive parts of the surface potential of a molecule, V_b and V_a, adiabatic potential V_-. Diabatic (α) and adiabatic (β) surface-gas transitions.

parts, a binding potential V_b and an anti-binding part V_a. In an adiabatic approximation, which is valid in such cases, the system can be considered relaxed every time the potential line V_- is valid.

These curves are generally fitted by $-A \cdot r^{-1}$ for the attractive part and $+B \cdot r^{-7}$ for the repulsive one, A being the Madelung constant times a factor larger than unity, and B is then evaluated with the equilibrium distance molecule - surface [6]. In an adiabatic case, as valid, e.g., for slow thermal heating, molecules will most probably decompose instead of gaining the desorption energy as a whole, because the decomposition energies are by far lower than the desorption energies of ions for all practical cases. However, in rapid processes, a non-adiabatic transition to V_a is possible prior to equilibration of energy. Thus, the intact molecule may gain the kinetic energy $E_{kin} = B \cdot r_o^{-7}$, which is reasonably large. The internal energy, however, is only due to the relaxation energy, leading to only slow decomposition of the molecule in the gas phase. This model has been tested quantitatively in the Al_2O_3 case by measurements of kinetic energy, IR measurements of the internal energy, and several other kinetic and thermodynamic features by DREYFUS et al. [6], and has thus been confirmed.

With this model the emission of ions should be qualitatively as well as quantitatively independent of the method of energy deposition, provided it is rapid enough to overcome thermal decomposition, and so strong, that sufficiently large surface area is affected by a minimum energy density around 2 kJ/cm^3, compared to a polarizing electric field strength of 10^8 V/cm. As SÄWE et al have shown [7], larger molecules are desorbed only if the area of sufficient excitation exceeds their magnitude. By the way, their "multi-hit" model is quantitatively identical for large m. Namely, in both models the emission yield Y is given by

$$Y = \frac{2 \pi P_s}{L^2} \int_0^{r_{max}} r \, P_{des} \, dr \qquad (1)$$

with P_{des} being in their model the Poissonian "multi-hit" probability for δ-rays in HIID, and P_{des} in our theoretical model being simply $P_{des} = 1$, if the surface high-frequency polarizing field exceeds the above value at the point r under consideration, otherwise $P_{des} = 0$. L is the linear dimension of the molecule, P_s a chemistry-dependent probability of desorbing the particular molecular ionic species, as discussed elsewhere [5], and r the distance, projected onto the surface, from the center of any concentric excitation whatsoever. With this formula (1), we may now easily compare the excitation functions for various methods.

3. Excitation Functions of Diabatic Transitions

Equation (1) can be rewritten as $Y = P_s \cdot A_{exc}/A_{mol}$ with A_{exc} the surface area of sufficient excitation and A_{mol} the area covered by one molecule of desorption probability P_s. As an example SÄWE et al. [7] find $P_s = 0.028$ for the quasimolecular cation of valine. We may compare this value with results of laser irradiation with a gaussian spatial density distribution of the UV-laser light. An energy density of about 0.5 to 1.0 J/cm^2 - depending on absorption properties - is found in our experiments, as well as in those of DREYFUS et al. [6], to be the threshold for ion formation from polar dielectrics. Within this desorption mode regime (about 10^8 W/cm^2 with 10 ns pulse length), the area of excitation determines the yield of the emitted ions. With the LAMMAR-1000 instrument $A_{exc} = 10^{-7}$ cm^2 can be assumed. A_{mol} (valine) is 3×10^{-13} cm^2. With (1) we obtain $Y = 10^4$ quasimolecular ions. This order of magnitude is found, indeed, in cases such that no macroscopic ablation of the irradiated surface material is produced.

With Electric Pulse Induced Desorption, as described in detail in Sect. 4, a 5 µm-⌀ wire surface is excited. Due to ion optics, only ions emitted in angles ± 1° relative to the axis are accepted by the spectrometer (radial electric field on the wire surface!), which is a 10^{-5} cm wide strip along the wire, whose active length is about 10^{-1} cm. Thus $A_{exc} = 10^{-6}$ cm^2. Homogeneous coverage provided, a number of 10^5 quasimolecular ions of valine is expected. This yield is found, indeed, with thin (tens of monolayers) amino acid layers, being prepared by an evaporation technique by HOLTKAMP and BENNINGHOVEN.

With dust particles [1] we calculated a ring-type active area with sufficient energy density around a, e.g., R = 5 µm dust particle impacting a surface electrosprayed with amino acids of a ring-diameter dR = 3 µm. Consequently $A_{exc} = 2\pi R dR = 10^{-6}$ cm^2, thus yielding nearly the same quantity of quasimolecular ions as found under EPID conditions.

From the point of view of excitation functions, as well as from that of all the other aspects as discussed elsewhere [5], we are sure that all these "fast" desorption techniques can be treated theoretically by the same diabatic model.

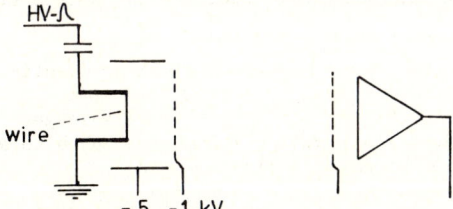

Fig. 2 Experimental set up of the EPID method

4. The EPID method

As shown in Fig. 2, a 2 ns electric pulse of about U = 1 kV excites a 5 µm-⌀ tungsten wire, which is covered by the dielectric under investigation. It lasts some 100 ns until the whole dielectric gains its maximum temperature, and some milliseconds for cooling down again. This has been measured by observing the glow light, using higher voltages up to 2 kV for this purpose. During this time period, thermionic emission is also observed. However, even with lower voltages applied, an instantaneous burst of ions, at least not longer than some 20 ns, is emitted from the dielectric, showing the same mass spectral behaviour well known from the other techniques mentioned. A convential time-of-flight mass spectrometer is used for detection.

However, due to the fact that the whole probe is affected thermally in every pulse event, no labile substances can be investigated. On the other hand, this very

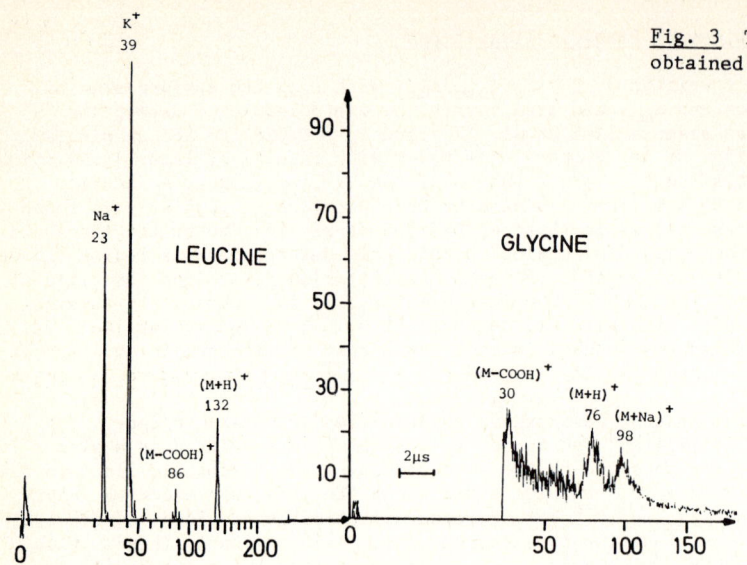

Fig. 3 Typical mass spectra obtained by the EPID method

simple method seems to be useful for some fundamental thermodynamic investigations of diabatic desorption. Two typical mass spectra of amino acids are reproduced in Fig. 3.

5. Applications of the EPID method

Dust impact sensors are on board the European Halley mission GIOTTO and the Soviet missions VEGA I+II. In order to calibrate the amplification of the ion detectors just before encounter, EPID units emit Cs^+ and Cs_2I^+ signals, which yield ratios serving to adjust the detector voltages. No other practical application has been achieved up to now. At least one may think about EPID as a simple detection technique for HPLC, as it detects polar organics with great sensitivity.

References

1. F.R. Krueger, W. Knabe, Organic Mass Spectrom. 18, 83 (1983)
2. P.K.D. Feigl, F.R. Krueger, B. Schueler, Organic Mass Spectrom. 18, 442 (1983)
3. R. Kelly, J.E. Rothenberg, Nucl. Instr. & Meth. in Phys. Res. B7/8, 755 (1985)
4. F.R. Krueger, Surface Science 86, 246 (1979)
5. F.R. Krueger, Z. f. Naturforschung 38a, 385 (1983)
6. R.W. Dreyfus, R.E. Walkup, R. Kelly, Radiation Effects (1985), in print
7. G. Säwe, A. Hedin, P. Hakansson, B. Sundqvist, DIET II, Springer, p. 213 (1985)

Part VI

Instrumentation

A New Dual-MS Technique Combining Negative Ion Formation by Plasma Desorption with EI-like Positive Ion Formation by In-Beam Desorption

H. Brandenberger and F.B.Ch. West

Department of Forensic Chemistry, University of Zürich,
Zürichbergstrasse 8, CH-8028 Zürich, Switzerland

Regardless of the recent development of many new ionization modes, EI-MS is still the method of choice for the characterization of low molecular weight compounds. For heat labile and very polar molecules, however, an additional soft ionization technique can be extremely useful, i.e. for finding the molecular mass.

In the past years, we have developed a new dual-MS-combination which furnishes, quasi-simultaneously in a single operation, conventional EI-spectra together with negative CI-spectra [1-3]. This combination is made possible since the pressure dependance of chemical ionization in the positive and in the negative mode is basically different (Fig. 1). In the positive mode, EI-fragmentation dominates up to a pressure of nearly 0.1 torr. It is hardly affected by the presence of reagent gas in the pressure range below. Pure positive CI-spectra require reagent gas pressures of 0.5 torr or higher. In the negative mode, EI-fragmentation is practically nonexistent. CI-spectra can be observed - in an open ion source - already at reagent gas pressures of slightly over 0.001 torr. Around 0.01 torr, the pressure dependance curve of negative CI shows a plateau. In this region, it is possible to obtain side by side positive EI- and low-pressure negative CI-spectra.

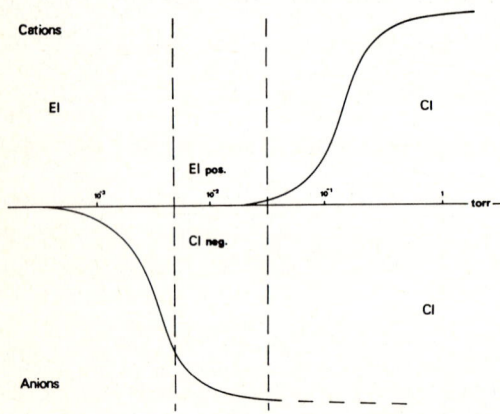

Fig. 1:
Pressure dependance of positive and negative ionization modes

There can be quite a difference between the negative ion spectra obtained at low and at high ion source pressure [4]. The lower the pressure, the more energy is transferred in the ionization step, the more fragmentation can occur. Therefore, our low-pressure negative CI-spectra usually show more fragment ions than medium-pressure CI-spectra (around 1 torr) or high-pressure negative CI-spectra (atmospheric pressure). We have found this extremely useful for structural elucidation, since the negative fragmentation scheme yields analytically significant information not inherent in the positive EI-fragmentation pattern. The two methods are complementary; their combination does not duplicate results, as is at least in part the case with the two previously described dual-MS-techniques: the combination of EI with positive CI, and the combination of positive and negative CI.

Fig.2 shows the format we have chosen for our dual-MS-presentation: the positive EI-spectrum as conventional bar diagram, the negative CI-spectrum upside down with the same mass scale as basis. The spectral pair permits to identify the compound as diethylene glycol. With only one of the 2 spectra, positive identification would hardly be possible.

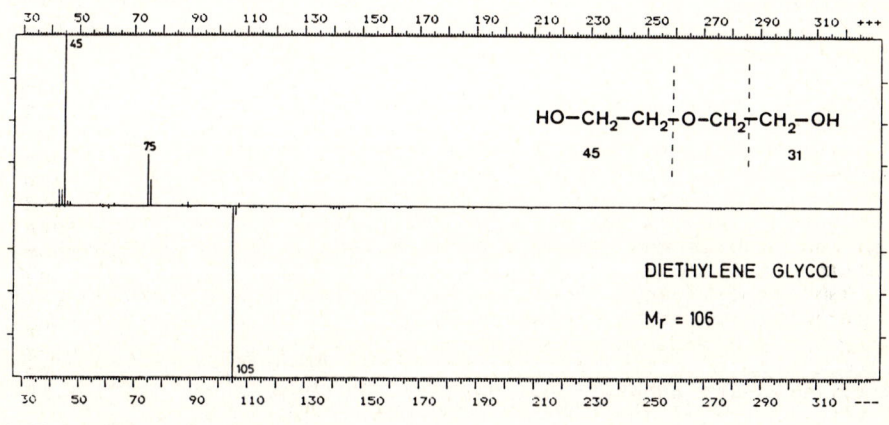

Fig.2: Quasi-simultaneous recording of positive EI- and negative CI-spectra of diethylene glycol by GC-MS.

The spectra were recorded with a quadrupole instrument R 10/10C from Nermag, capable of switching from positive to negative ion recording and vice-versa in less than a second. Every uneven scan yields a positive EI-spectrum, every even scan a negative low-pressure CI-spectrum. Since there is an energy difference between the positive and negative ions formed, they must be adjusted separately at the entrance of the quadrupole by 3 pairs of potentiometers for focussing voltage, ion energy and extraction voltage in positive and negative. The negative ions are converted to positive ions by a separate conversion dynode. Amplification of both ion currents is effected by the same 21-step electron multiplier. The ion source is an open design, a slightly modified EI-source with a 5x1 mm slit for the filament electrons and a 4 mm diameter hole as ion exit, but without trap hole. The reagent gas can be added coaxial with the GC-inlet or perpendicular through a separate line.

The nature of the reagent gas (together with the source pressure) determines the negative ionization mechanism [5, 6]:

Conventional reagents such as CH_4, but also N_2 or noble gases, yield mainly positive reagent ions (for positive CI). In addition, they can be used as moderators for the production of low-energy electrons which can ionize sample molecules by electron attachment. According to the compound class, the resulting negative molecular ion is either stable or dissociates, usually by loss of H.

Some selected reagents yield - on electron impact - considerable concentrations of negative reagent ions which can ionize sample molecules by negative charge exchange (which may also be followed by dissociation), by proton abstraction, by substituting H or another substituent of the sample molecule, or by cluster anion formation.

While cluster ion formation, in the positive and in the negative mode, and to some extent also substitution reactions, are favored by high source pressure, electron attachment, negative charge exchange and proton abstraction work equally well at low pressure.

The reagent gases we use most often are:
- CH_4 for ionizing - by electron attachment - compounds with high electron affinity;
- N_2O - which yields O^- and $[NO]^-$ on electron impact - for ionizing compounds with low electron affinity by charge exchange or proton abstraction;
- CO_2 - which yields O^- and a considerable concentration of low energy electrons - for the ionization of unknown compounds.

We have now extended our dual-MS-combination to the characterization of heat labile and strongly polar molecules by in-beam desorption. This permits to obtain quasi-simultaneously positive EI-like desorption spectra and negative DCI-spectra. A few examples will illustrate the usefulness of the method.

Polyfunctional acids are difficult to identify from their EI-fragments only. The negative DCI-spectra indicate their mass by the intensive ions $[M-H]^-$ and $[2M-H]^-$ (Fig. 3).

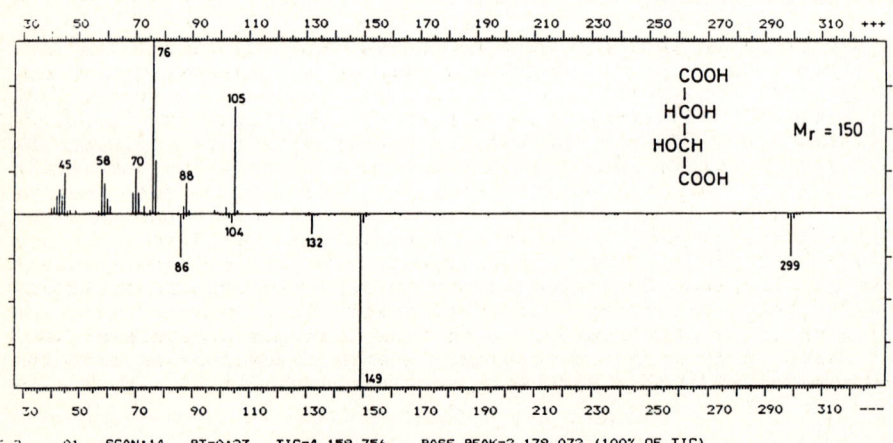

Fig. 3: Dual-MS-recording of a polyfunctional acid by in-beam desorption

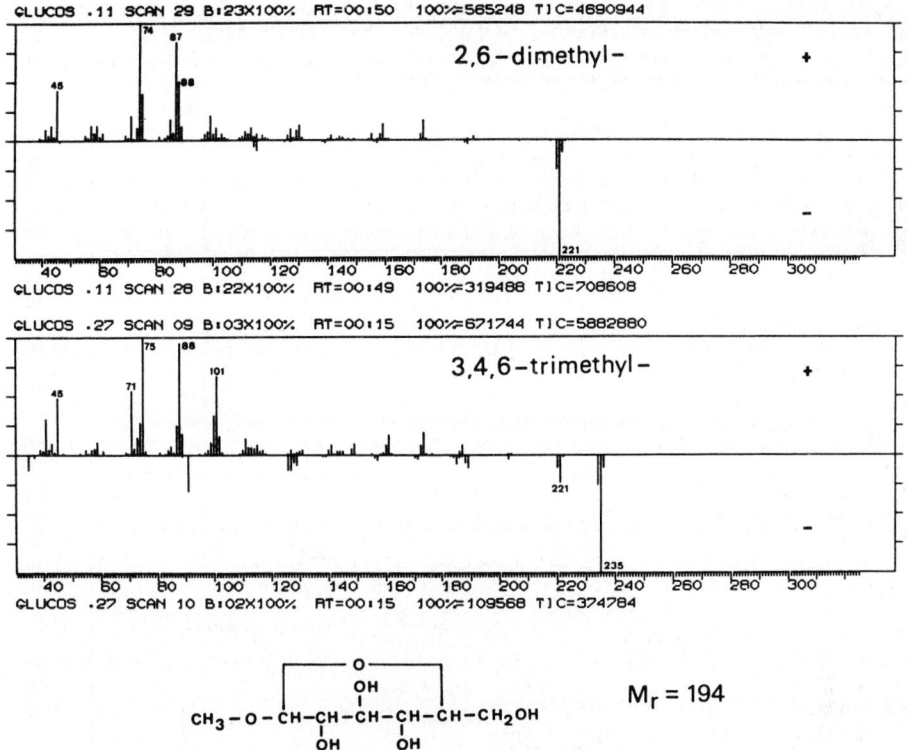

Fig. 4: Dual-MS-recording of 2 methylated methyl-glucosides by in-beam desorption

Methyl-glucosides can also be quite well characterized by our dual-MS-approach [7]. Figure 4 shows the spectra of two methylated compounds. The negative DCI-recordings indicate the mass and therefore the number of methyl groups. From the positive recordings, conclusions regarding the position of the methyl groups may be drawn (masses 45, 74/75, 87/88, 101).

The negative DCI-spectra of flavonoid-glycosides such as Rutin (Fig. 5) reveal the masses of the glycoside and of the corresponding aglycon [8]. The positive desorption spectra show, in addition to the aglycon moiety, some low-mass ions originating from the sugar residues.

Prostaglandins and low molecular weight peptides have also been characterized by our dual-MS-technique, which is very simple to apply: The substance in organic or aqueous solution is coated on a tungsten wire, the solvent evaporated and the wire introduced into the center of the ion chamber, with a reagent gas pressure close to 0.005 torr. The wire is heated in about 6 seconds from ambient temperature to around $1000°C$ with dual-MS-recording. Due to the low source pressure, contamination is negligible.

We recommend the technique especially for compounds yielding analytically useful EI-fragments without M_r-information. The dual-MS-recordings may also be helpful in obtaining information about the ionization mechanism in DCI.

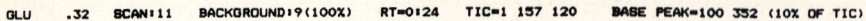

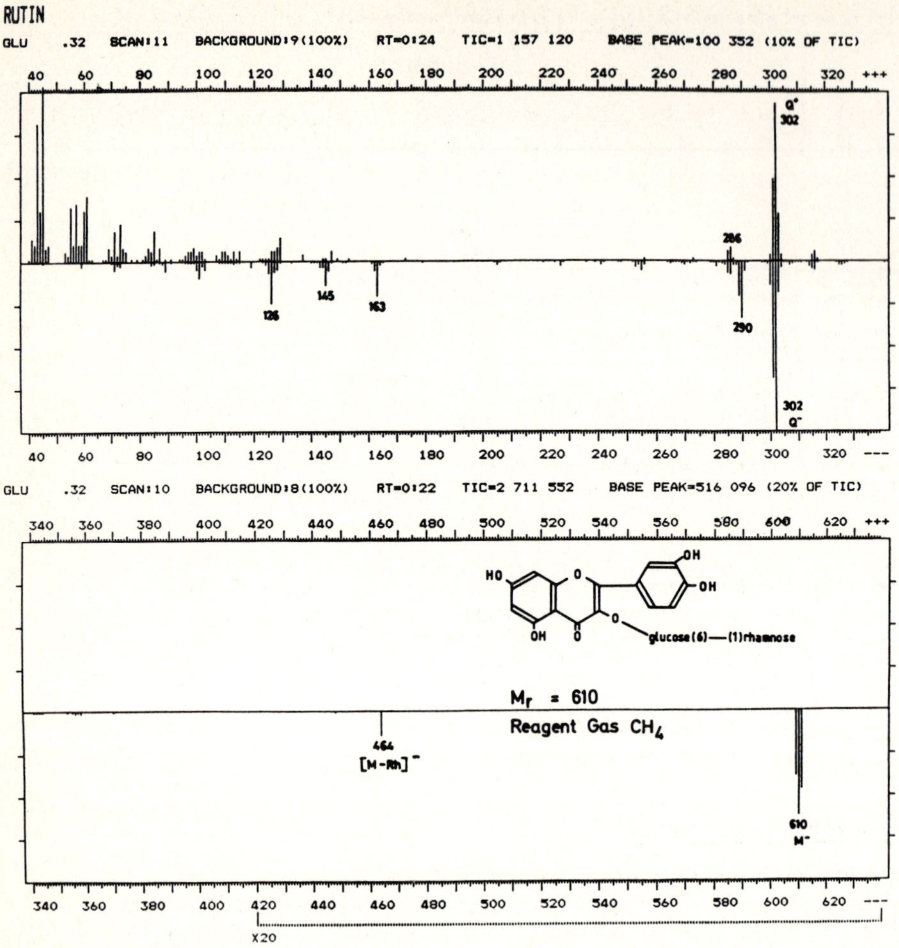

Fig. 5: Dual-MS-recording of a flavonoid glycoside by in-beam desorption

References

1. H. Brandenberger: 30th Ann. Conference on Mass Spectrometry and Allied Topics, June 1982, Honolulu, abstract volume p. 468.
2. H. Brandenberger: Int.J.Mass Spectrom. Ion Phys. 47, 213 (1983).
3. F.B.Ch. West and H. Brandenberger, in R.A.A. Maes (Ed.): Topics in Forensic and Analytical Toxicology (Elsevier, Amsterdam 1984).
4. H. Brandenberger, in A. Frigerio and M. McCamish (Ed.): Recent Developments in Mass Spectrometry in Biochemistry and Medicine 6, p. 391 (Elsevier, Amsterdam 1980).
5. H. Brandenberger und R. Ryhage, in A. Frigerio (Ed.): Recent Developments in Mass Spectrometry in Biochemistry and Medicine 1, p. 327 (Plenum, New York 1978).
6. H. Brandenberger, in A. Frigerio (Ed.): Recent Developments in Mass Spectrometry in Biochemistry and Medicine 2, p. 227 (Plenum, New York 1979).
7. A. Hayashi, F.B.Ch. West and H. Brandenberger: to be published.
8. A. Sakushima, F.B.Ch. West and H. Brandenberger: to be published.

The Chemical Ionization/Particle-Induced Ion Source

R.B. Freas and J.E. Campana

Naval Research Laboratory, Chemistry Division, Washington, DC 20375, USA

1. Introduction

Fast-atom bombardment (FAB) has proven to be a useful desorption technique for the mass spectrometric analysis of intractable chemical species [1]. These species are sputtered as ions and neutrals from solid surfaces, films, or liquid matrices by an energetic beam of fast atoms. The number of sputtered neutral species should be considerably larger than the number of ionic species, based on typical ion yields from classical surface studies.

A high-pressure, fast-atom bombardment (FAB) ion source has been constructed to study ion/molecule reactions of sputtered ions and neutrals [2,3]. Pressures in this ion source are typically on the order of 0.1 to 0.5 Torr. Studies, involving the fundamentals of ion/molecule reactions of desorbed organic species and metal cluster chemistry, have been performed using this chemical ionization/fast-atom bombardment (CI/FAB) source with the mass-analyzed ion kinetic energy spectrometry (MIKES) technique. Our experiments were performed using a reverse-geometry, double-focusing mass spectrometer. Xenon gas was used as the fast-atom beam.

2. Collisional Stabilization

Collisional stabilization of sputtered ions was affected by different buffer gases, He, Ar, Xe, CH_4, CO, CO_2, and $i-C_4H_{10}$. The unimolecular dissociation of the protonated molecule of pentaerythritol tetrapentanoate (PETP) was observed to decrease in direct proportion to the effective collisional stabilization cross-section of the buffer gas (Fig.1).

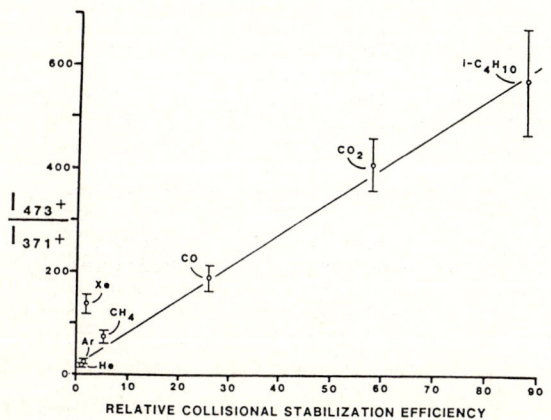

Fig. 1: The ratio of the abundance of the protonated molecule of PETP ([M+H]$^+$, m/z 473) to the unimolecular dissociation product ion (loss of an acid moiety, m/z 371) versus the relative collisional stabilizational efficiencies of several gases. The plot provides a good correlation between the relative ion abundance enhancements and the collisional stabilization efficiencies of the several different gases.

3. Post-desorption Ionization

Post-desorption ionization (PDI) of sputtered neutrals is accomplished by ion/molecule reactions through the generation of reactant ions in a conventional manner. The PDI of sputtered neutrals has been observed to increase the abundance of an [M+H]$^+$ species by almost three orders of magnitude. Processes occurring within the ion source include proton transfer, charge exchange, and association reactions [3].

Ion abundances of the [M+H]$^+$ species of PETP and fragment ions change with the addition of a buffer gas to the ion source. Figure 2 displays mass spectra of PETP with different bath gases. The use of NH$_3$ as a bath gas collisionally stabilizes the molecular ion and results in the formation of an [M+NH$_4$]$^+$ species. This association reaction has permitted the analysis of the composition of a number of different pentaerythritol tetraester isomers in a commercial lubricant formulation.

Figure 3 displays the negative ion mass spectra of PETP under conventional FAB and CI/FAB conditions. The major ions that are observed in the FAB negative ion spectrum (Fig. 3a) are the [M-H]$^-$ (m/z 471) species and the

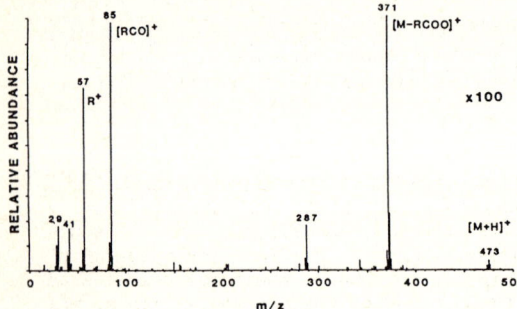

Fig. 2a: A fast-atom bombardment mass spectrum of neat PETP obtained in the CI/FAB ion source with no reactant gas or ionizing electrons. The protonated molecule ([M+H]$^+$, m/z 473) is less than 5% of the base peak, [M-RCOO]$^+$. The protonated molecule is less than 0.5% of the base peak in the fast-atom bombardment mass spectra obtained in a conventional FAB ion source [2].

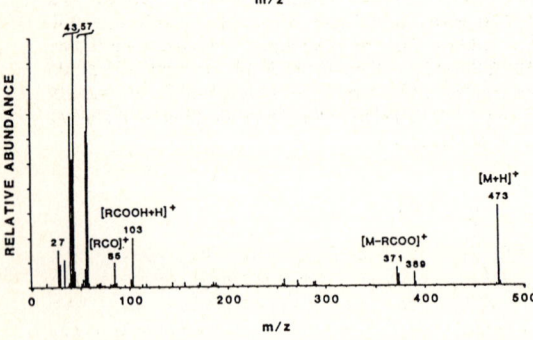

Fig. 2b: An isobutane chemical ionization/fast-atom bombardment mass spectrum of neat PETP. The isobutane pressure was 0.2 Torr. The molecular ion ([M+H]$^+$, m/z 473) is almost 1000 times more abundant than in the FAB spectrum in Fig. 2a. The [M+H]$^+$ species is more abundant than any of the fragment ions of PETP.

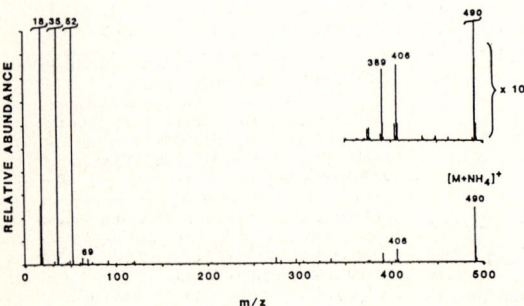

Fig. 2c: An ammonia chemical ionization/fast-atom bombardment mass spectrum of neat PETP. The molecular adduct ion [M+NH$_4$]$^+$ is almost 1000 times more abundant than the [M+H]$^+$ species in the FAB spectrum shown in Fig. 2a. The ammonia pressure was 0.2 Torr. High-mass fragment ions are observed analogous to those seen in the isobutane CI/FAB spectrum of PETP.

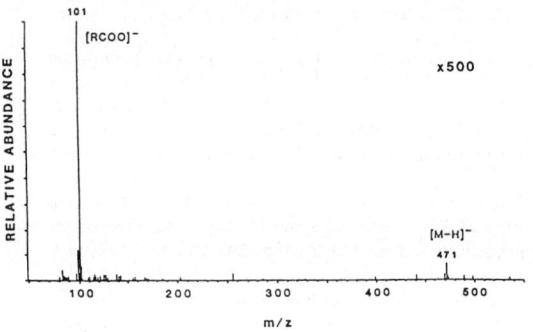

Fig. 3a: A negative ion fast-atom bombardment mass spectrum of neat PETP.

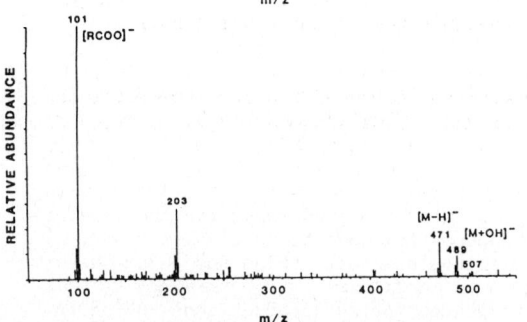

Fig. 3b: An H_2/N_2O $[OH]^-$ negative ion chemical ionization/fast-atom bombardment mass spectra of neat PETP. The deprotonated molecule $[M-H]^-$ (m/z 471) is almost 1000 times more abundant than in the negative ion FAB spectrum shown in Fig. 3a. The H_2/N_2O pressure was 0.2 Torr.

carboxylate negative ion of the C_5 acid moiety, ($[C_4H_9CO_2]^-$, m/z 101). Figure 3b shows the CI/FAB mass spectrum that was obtained with a mixture of H_2 and N_2O (1:1) as the reagent gas mixture with the ionizing filament on. This gas mixture yields the $[OH]^-$ reagent species. The $[OH]^-$ CI/FAB mass spectrum of PETP gives an almost 1000-fold enhancement of the $[M-H]^-$ abundance relative to the spectrum shown in Figure 3a. The $[M+OH]^-$ adduct species and an ion at m/z 203 are observed also in the $[OH]^-$ CI/FAB mass spectrum of PETP, which are not present in the FAB mass spectrum. This latter ion corresponds to the negative ion dimer of the fatty acid moiety $[(RCOOH)_2-H]^-$ from the PETP molecule, which may be written equivalently as the proton-bound dimer of the carboxylate anions ($RCOO^-...H^+...^-OOCR$). Ionic clusters such as these have been observed also in the CI/FAB spectra of pure fatty acids and esters [4].

4. Formation and Reactions of Metal Cluster Ions

The fast-atom bombardment of metal foils produces sputtered metal cluster ions M_x^+ (Reaction 1). When the desorption occurs within the high-pressure ion source, the bare metal cluster ions react with gas-phase species to form unique metal cluster product ions (Reaction 2). The neutral reactant species can be any volatile species, such as organic molecules (alkanes, alcohols, or alkyl halides), inorganic species (NH_3, O_2, or CO), or organometallic compounds. The reactions of the bare metal cluster ions (Reaction 2) may be analogous to the sorption of the reactant species on metal surfaces.

$$M_{(s)} + Xe \text{ (keV)} \longrightarrow M_x^+ \qquad (1)$$

$$M_x^+ + yR \longrightarrow [M_xR_y]^{+*} \longrightarrow \text{Products} \qquad (2)$$

When oxygen is admitted to the ion source, iron and cobalt cluster ions react to produce abundant metal/oxygen cluster ions $[M_xO_{x-1}]^+$, $[M_xO_x]^+$, and low abundances of $[M_xO_{x+1}]^+$. The $[M_xO_y]^+$ cluster ions were not detected by sputtering of the target foil after the O_2 was removed from the ion source.

The $[M_xO_y]^+$ cluster ions were observed by sputtering cobalt oxide. These results suggest that oxygen is not chemisorbed extensively on the metal surface, and thus is an indication that the oxygenated cluster ions were formed in the gas-phase.

5. Structure and Bonding of Metal Cluster Product Ions

In an experiment to probe structure and bonding using tandem mass spectrometry, a mass-selected product ion is dissociated (unimolecular or bimolecular dissociation, Reaction 3) and the fragment ions are energy analyzed.

$$[M_xR_y]^+ \xrightarrow{\text{Dissociation}} [M_iA_j]^+ \quad + \quad [M_kB_l]^+ \quad + \ldots \qquad (3)$$

The collision-induced dissociation (CID) spectra of metal/oxygen cluster product ions show major fragmentations that correspond to losses of M-O units or formation of $[M_iO_i]^+$ fragment ions. These pathways indicate the formation of stable structures of stoichiometry $[Co_4O_4]^+$, $[Co_3O_3]^+$ and $[Co_2O]^+$.

The different fragmentation pathways of the $[Co_xO_x]^+$ and $[Co_xO_{x-1}]^+$ species in Tables 1 and 2 are indicative of structural differences in the cluster ions related to their stoichiometry. We are inferring structures for the cluster ions based upon the observed collision-induced fragmentations (Tables 1 and 2) and an ionic model (consisting of an electrostatic attraction and a Born-Mayer repulsion [5]). The fragmentations of the ionic models by cleavage of the fewest bonds yield the most stable structures for the product ions and correspond to observed CID pathways [6].

Table 1. Percent abundances of fragments from the CID of $[Co_xO_x]^+$

			Neutral Fragment					
Parent Ion	Co	CoO	CoO_2	Co_2O_2	Co_2O_3	Co_3O_3	Co_3O_4	Co_4O_5
$[Co_5O_5]^+$	-	100	8	64	8	78	31	18
$[Co_4O_4]^+$	-	100	12	25	33	4	15	
$[Co_3O_3]^+$	18	85	100	12	59			
$[Co_2O_2]^+$	56	15	100					

Table 2. Percent abundances of fragments from the CID of $[Co_xO_{x-1}]^+$

			Neutral Fragment			
Parent Ion	Co	CoO	Co_2O	Co_2O_2	Co_3O_3	Co_3O_4
$[Co_5O_4]^+$	100	-	79	-	79	45
$[Co_4O_3]^+$	100	29	-	94	71	
$[Co_3O_2]^+$	48	100	4	52		
$[Co_2O]^+$	11	100				

Isobutane and Fe, Co, Cu, and Ag cluster ions react to form metal/isobutane product ions, $[M_x(C_4H_{10})_y]^+$ (x=1-5; y=1-3). The CID fragment ions of cobalt/isobutane cluster ions in Table 3 indicate the bonding interactions between the metal atom(s) and the isobutane C-C or C-H bonds. The cobalt ion Co^+ activates C-C and C-H bonds (exothermic bimolecular reactions [7]), and this is also observed in the CID of $[CoC_4H_{10}]^+$. The dicobalt ion Co_2^+ does not react exothermically [7], and the CID of the $[CoC_4H_{10}]^+$ species shows no activation of hydrocarbon bonds. In contrast to Co_2^+, $[Co_3C_4H_{10}]^+$ interacts with C-H bonds as observed from the CID loss of H_2.

Table 3. Percent abundances of fragment ions from the CID of $[Co_xC_4H_{10}]^+$

$[CoC_4H_{10}]^+$		$[Co_2C_4H_{10}]^+$		$[Co_3C_4H_{10}]^+$	
Fragment	%	Fragment	%	Fragment	%
$[CoC_4H_9]^+$	2	Co_2^+	100	$[Co_3C_4H_8]^+$	10
$[CoC_4H_8]^+$	33	$[CoC_4H_8]^+$	2	Co_3^+	90
$[CoC_4H_7]^+$	2	$[CoC_3H_6]^+$	2	$[Co_2C_4H_8]^+$	21
$[CoC_3H_6]^+$	100	$[CoCH_3]^+$	1	$[Co_2H]^+$	86
$[CoCH_3]^+$	8	Co^+	14	Co_2^+	100
Co^+	55			Co^+	21

The $[Co_xO_{x-1}]^+$ clusters react with isobutane in the CI/FAB ion source to form $[Co_xO_{x-1}C_4H_8]^+$ as well as $[Co_xO_{x-1}C_4H_{10}]^+$ products. The CID fragmentations show the abundant loss of H_2 from $[Co_2OC_4H_{10}]^+$ (Table 4). The $[Co_2O_2C_4H_{10}]^+$ cluster does not lose H_2 by CID to give $[Co_2O_2C_4H_8]^+$ (Table 4), and $[Co_2O_2C_4H_8]^+$ was not observed in the CI/FAB spectrum. Therefore, the oxygen-deficient cobalt clusters $[Co_xO_{x-1}]^+$ dehydrogenate isobutane, but the $[Co_xO_x]^+$ clusters do not.

Table 4. Percent abundances of fragments from the CID of $[Co_xO_yC_4H_{10}]^+$

Parent Ion	Neutral Fragment					
	H	H_2	O	H_2O	C_4H_8	C_4H_{10}
$[CoC_4H_{10}]^+$	2	33			–	55
$[CoC_4H_8]^+$	3	15			100	
$[Co_2OC_4H_{10}]^+$	–	90	10		–	100
$[Co_2OC_4H_8]^+$	3	3	12	11	100	
$[Co_2O_2C_4H_{10}]^+$	5	2		2	–	100
$[Co_2O_2C_4H_8]^+$			Not Observed			

Thus, CI/FAB mass spectrometry provides a tool for the study of ion/molecule reactions of sputtered species. Tandem mass spectrometry provides a probe of the structure and bonding of metal cluster product ions. In addition, the cluster reaction product ions are inherently ions; they do not need to be photoionized as in molecular beam studies, and hence are not susceptible to perturbations upon ionization.

6. References

1. M. Barber, R.S. Bordoli, G.J. Elliott, R.D. Sedgwick, A.N. Tyler: Anal. Chem., 54, 645A (1982)
2. J.E. Campana and R.B. Freas: J. Chem. Soc. Chem. Commun. 1414 (1984)
3. R.B. Freas and J.E. Campana: J. Am. Chem. Soc. in press
4. M.M.Ross, R.A. Neihof, J.E. Campana: submitted for publication
5. T.P. Martin: J. Chem. Phys. 72, 3506 (1980)
6. R.B. Freas, B.I. Dunlap, and J.E. Campana: unpublished results
7. R.B. Freas and D.P. Ridge: J. Am. Chem. Soc. 102, 7129 (1980)

Design of Modern Time-of-Flight Mass Spectrometers

H. Wollnik

II. Physikalisches Institut, Universität Giessen,
D-6300 Giessen, F.R.G.

Abstract

The performance of time-of-flight and laterally dispersive mass spectrometers is compared. Figures of merit of both systems are given, and fundamental as well as technical limits of resolving power and transmission are discussed.

1. Introduction

The velocity of low-energy ions is determined as $v=\sqrt{2K/m}$ with m being the mass and K the kinetic energy of these ions. With K in MeV and m in mass units u the magnitude of v in m/μsec is found as

$$v \approx 14\sqrt{K/m} \quad . \tag{1}$$

To make the flight time of ions depend only on the ion mass, the optical system under consideration should be designed such that ions of higher energy are sent on a detour, the length of which increases linearly with the ion energy. Such designs require deflecting electrostatic and/or magnetic fields. In such properly designed ~~such~~ systems, the ion flight time is independent of the energy to charge ratio K/(ze), so that the ion flight time is a sole function of the mass-to-charge ratio m/(ze).

One such time-of-flight mass spectrometer is a homogeneous magnetic field of flux density B_0 in which the ions move on a radius ρ and their flight time through an angle o is

$$T = \frac{2\phi\rho}{\sqrt{2K/m}} = \frac{2\phi m}{B_0(ze)} \tag{2}$$

because of $(ze)vB_0 = mv^2/\rho$ or $\rho = \sqrt{2Km}/[(ze)B_0]$ (see also Fig.1). For ions of equal mass-to-charge ratio, thus the ion flight time is thus independent of the kinetic energy K, a fact which was recognized by E. Lawrence and used for the design of the cyclotron. A similar time-of-flight mass spectrometer can be built by using a clystron-like (see Fig.2) ion reflector [1]. The flight time in the field-free region of length ℓ_1 here is $T_1 = \ell_1/v = \ell_1\sqrt{m/2K}$ and the flight time in the repeller field of strength E is $T_2 = dz/v = 2\ell_2\sqrt{m/2K}$ where $\ell_2 E = K(ze)$ is the potential at which the ions of energy K come to rest. The total flight time is thus

$$T = 2[T_1 + T_2] = \sqrt{2m}\left(\frac{\ell_1}{\sqrt{K}} + 2\ell_2\sqrt{K}\right) \tag{3}$$

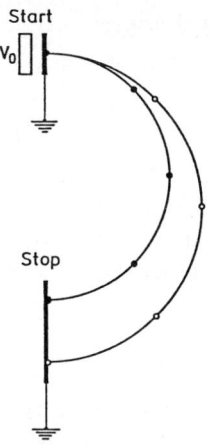
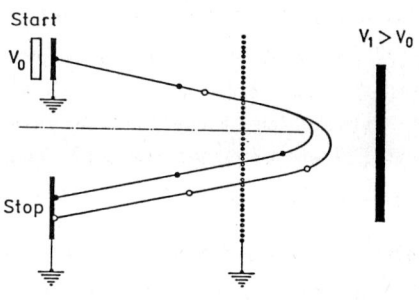

Fig. 1:
The motion of two ions of equal mass-to-charge but different energy-to-charge ratios in a homogeneous magnetic field. Note that the ions of higher energy, i.e. of higher velocities, are sent on a detour which has such a length that both ions arrive at the same time at the stop detector after a deflection of 180°.

Fig. 2:
The motion of two ions of equal mass-to-charge but different energy-to-charge ratios in a reflector-like time-of-flight mass spectrometer. Note that the ions of higher energy, i.e. of higher velocities, are sent on a detour which has such a length that both ions arrive at the same time at the stop detector.

with ion energies $K=K_0(1+\delta_K)$ this flight time T becomes to first order independent of δ_K for

$$\ell_1 = 2\ell_2 \; . \tag{4}$$

Also in this system the more energetic ions move on a detour since they penetrate deeper into the repeller field.

2. Laterally and Longitudinally Dispersive Mass Spectrometers

In an optical system like the magnetic sector field of Fig.3 the position x,y and inclination $a=p_x/p_0 \approx \sin\alpha$ and $b=p_y/p_0 \approx \sin\beta$ as well as the flight time difference τ of an arbitrary (K,m,ze) and a reference particle $(K_0, m_0, z_0 e)$ are generally described by

$$\begin{pmatrix} x_4 \\ a_4 \\ t_4 \\ \delta_K \\ \delta_m \end{pmatrix} = \begin{pmatrix} (x,x) & (x,a) & 0 & (x,\delta_K) & (x,\delta_m) \\ (a,x) & (a,a) & 0 & (a,\delta_K) & (a,\delta_m) \\ (t,x) & (t,a) & 1 & (t,\delta_K) & (t,\delta_m) \\ 0 & 0 & 0 & 1 & 0 \\ 0 & 0 & 0 & 0 & 1 \end{pmatrix} \begin{pmatrix} x_1 \\ a_1 \\ t_1 \\ \delta_K \\ \delta_m \end{pmatrix} \tag{5a}$$

$$\begin{pmatrix} y_4 \\ b_4 \end{pmatrix} = \begin{pmatrix} (y,y) & (y,b) \\ (b,y) & (b,b) \end{pmatrix} \begin{pmatrix} y_1 \\ b_1 \end{pmatrix} \; . \tag{5b}$$

Classical mass spectrometers use the lateral dispersion (x,δ_m) of magnetic sector fields in which an ion collector is placed at the focus position, i.e. at that position at which

$$(x,a) = 0 . \tag{6a}$$

ASTON [2] had the idea to combine magnetic and electrostatic sector fields so that (x,δ_K) would vanish and the mass dispersion become independent of the energy distribution of the ions. An even better idea was developed by BARTKY and DEMPSTER [3] who combined the angle and energy focusing properties so that:

$$(x,a) = (x,\delta_K) = 0 . \tag{6b}$$

Very generally, the performance of laterally dispersive systems is described [4] by the $Q_{x,K}$- and $Q_{x,m}$-values:

$$Q_{x,K} = (2x_{10}2a_{10})\frac{1}{\delta_{K,min}} = \sum_i \frac{A_{aBi}}{2\rho_{Bi}} + \sum_i \frac{A_{aEj}}{\rho_{Ej}} \tag{7a}$$

$$Q_{x,m} = (2x_{10}2a_{10})\frac{1}{\delta_{m,min}} = \sum_i \frac{A_{aBi}}{2\rho_{Bi}} \tag{7b}$$

(see also Fig.3). Here ρ_{Bi} and ρ_{Ej} are the radii of curvature of the optic axes in the i_{th} magnetic and the j_{th} electrostatic sector field, while A_{Bi} and A_{Ej} denote corresponding areas in these sector fields encompassed by trajectories which diverge from the center of the object. Furthermore, $2x_{10}$ is the width of the entrance slit and $2a_{10}$ characterizes the maximal aperture angle $\pm a_{10}$ with $x_{10}a_{10}$ being the laterally accepted phase space, a measure for the ion beam intensity. Finally $\delta_{K,min}$ and $\delta_{m,min}$ describe the minimally resolvable energy

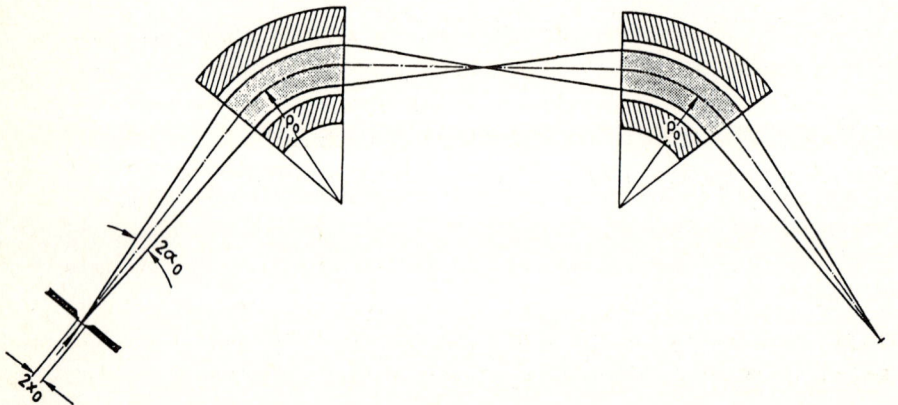

Fig. 3:
A time-of-flight mass spectrometer in which the ion flight time depends on the mass-to-charge ratio but is independent of the energy-to-charge ratio. As compared to the system of Fig.1, here a stigmatic achromatic focusing is achieved. Note that the two shaded areas encompassed in the two electrostatic sector fields by trajectories starting from the center of the object under angles $\pm a_0$ are equal in size. Thus the lateral energy dispersion (x,δ_K) vanishes as long as the system is laterally focusing, i.e. $(x,a)=0$.

and mass differences. Note here that the Eqs. 7 do not change if inhomogeneous sector fields or inclined magnet boundaries are employed, and that the use of Einzel- or quadrupole-lenses does not change Eqs. 6 either.

Postulating zero energy dispersion is equivalent to $K/\delta_{K,min}=0$ in Eq. 7a. For a system consisting of one magnetic and one electrostatic sector field, thus Eq. 7a yields:

$$A_B/\rho_B = 2A_E/\rho_E .$$

This leaves Eq. 7b unchanged. For more complex mass spectrometers, Eq. 7a describes the necessary relation between the electrostatic and the magnetic sector fields and Eq. 7b yields the achievable mass resolving power $m_0/\delta_{m,min}$. Thus a simple inspection of the areas A_{Bi}, A_{Ej} and radii ρ_{Bi}, ρ_{Ej} allows to judge the performance of a roughly designed optical system.

Time-of-flight mass spectrometers make use of the longitudinal dispersion $(t,\delta_m)\delta_m$. Such systems are designed [5] to fulfill

$$(t,\delta_K) = 0 \tag{8a}$$

in which case the flight time becomes independent of the energy distribution of the ions. Additionally to Eq. 8a one normally postulates achromatic and stigmatic lateral focusing, i.e.

$$(x,\delta_K) = (x,a) = (y,b) = 0 \tag{8b}$$

which is achieved by an appropriate geometry of the optical system. Having fulfilled Eq. 8b one finds [6] automatically $(t,a)=0$ so that $t_4=(t,x)x_1+(t,\delta_m)\delta_m$ according to Eq. 5a. As long as x_1 is small there is no major deterioration of the time-of-flight mass resolving power. However, one also can design the time-of-flight mass spectrometer such that (t,x) vanishes.

For a time-of-flight mass spectrometer in which

$$(x,a) = (t,\delta_K) = 0$$

one can derive a quality factor or $Q_{t,m}$-value [6] analogously to Eqs. 7:

$$Q_{t,m} = (t,\delta_m)2\delta_m = \sum_i \frac{A_{Ki}}{\rho_i} \tag{9}$$

(see also Fig.4). Here ρ_i denotes the radius of curvature of the optic axis in the i_{th} sector field and A_{Ki} the corresponding areas in these sector fields encompassed by trajectories of particles of the energy-to-charge ratios

$$K/(ze) = K_0(1\pm\delta_K)/(z_0 e).$$

Having obtained a special design, the $Q_{t,m}$-value also allows an easy judgement of its performance. As an example, we may inspect the system of Fig.4 and find that mainly the second and third sector fields contribute to the performance of the time-of-flight system, while the first and fourth sector fields only match the system to the ion source and the ion detector. Though Eq. 9 was derived for

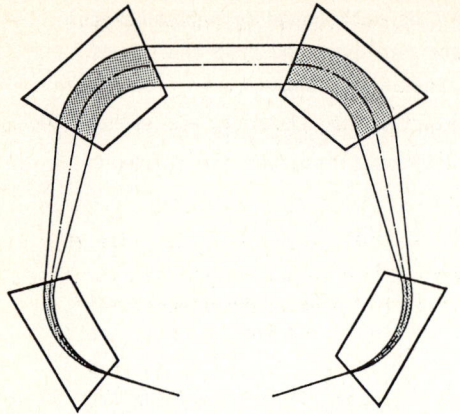

Fig. 4:
A time-of-flight mass spectrometer the ion flight time through which depends on the mass-to-charge ratio but is independent of the energy-to-charge ratio. The achievable mass resolving power is characterized by the shaded areas encompassed in the two sector fields by trajectories of ions of different energy-to-charge ratios, where these ions are assumed to start from the center of the object under zero angles.

sector field time-of-flight mass spectrometers, it applies as well to clystron-like reflectors where the area A_K is encompassed by the two trajectories of Fig.2 and the radius ρ varies with the ion energy.

3. Ion Sources and Detectors in Time-of-Flight Mass Spectrometers

In laterally dispersive mass spectrometers, ion sources and detectors influence only the recordable ion intensity as well as the ease of use of the spectrometer. However, they leave the mass resolving power of the system unchanged. Longitudinally dispersive mass spectrometers, on the other hand, use the ion source and detector as integral parts. The achievable mass resolving power thus directly reflects the performance of the time-of-flight spectrometer as well as the ion source and the ion detector.

Originally [2,8] all laterally dispersive mass spectroscopes were built as spectrographs with a photographic plate as recording instrument. With improved electronics the photographic plate was replaced by ion recorders [9] behind one or two exit slits using the mass spectroscope as a mass spectrometer. This had the advantage of instantaneous beam recording. However, it allowed one to record at a given time only one mass line and not a full spectrum, which was a great loss in the overall recorded ion intensity per time unit. In most longitudinally dispersive mass spectrometers all ions generated in one ionization pulse can easily be recorded, so that one always has a mass spectrograph. The problem is that - except for very complex systems - the time resolution of any ion counter is about 1 nsec for the time being, so that about 3 nsec are necessary for a positive identification of a mass line. To achieve a mass resolving power of 10000 or 100000, flight times of at least 30 to 50 μsec or 300 to 500 μsec are thus required.

Ion sources suitable for time-of-flight mass spectrometers must produce ions in a pulsed mode. For this type of ion generation the sample is advantageously prepared as a solid or possibly as the surface of a liquid matrix (for instance, glycerol). To release ions of interest, this surface is then bombarded by a laser [10] or a primary ion pulse [11] or by a short electrical pulse [12] passing through the substrate. The actual ion formation seems to take place in 10^{-12} sec or less, thus it is a question of short pulse generation to define the precisely the start time of the ions of interest.

In case of lasers, times below 1 nsec are well feasable. To obtain 10000 primary ions in a pulse of 1 nsec the average ion current must be about 1 μA. Since this is a reasonable high ion current, one often uses only 0.01 μA average ion currents, uses pulsed ion deflectors to cut pulses of about 100 nsec, and compresses those pulses to one or several nsec.

Ions of mass 100 and energy 10 keV move with velocities of about 1.4 cm/μsec. An ion pulse of 300 nsec length is then 4.2 cm long with a diameter of usually several 0.1 cm. The above-mentioned compression can now be achieved in two ways. One possibility is to enter the pencil-like ion cloud into the region between two electrodes and very quickly apply potentials to these electrodes such that the ion cloud moves sideways, and thus in this direction has only a length of perhaps 0.3 cm or about 2 nsec. Another possibility is to use the clystron-like buncher technique, in which the head ions of the cloud are slightly decelerated and the tail ions are slightly accelerated, so that all ions arrive at the same time, a certain distance downstream.

In case of explicitly prepared solid samples or samples in a glycerol matrix pulsed laser and SIMS techniques are applicable. In case of gaseous samples pulsed laser techniques are feasable. Thus, only in the latter case can gaschromatographs easily be coupled to a time-of-flight mass spectrometer, while principally both methods can be used in case of a liquid chromatograph, though up to now only off-line coupling techniques have been used to any extent.

4. Principle Limits to the Performance of Time-of Flight Mass Spectrometers

The Eqs. 5 can be extended to also contain higher order elements. Table I shows in three columns the elements of the first three lines of Eq. 5a evaluated for the system of Fig.4 to second order where all elements which include δ_m are omitted. Naturally, all these matrix elements are obtained from the solution of the equations of motion. One way to describe these equations of motion is to use the canonical equations

$$\begin{pmatrix} \partial x/\partial T \\ \partial p_x/\partial T \\ \partial y/\partial T \\ \partial p_y/\partial T \\ \partial z/\partial T \\ \partial p_z/\partial T \end{pmatrix} = \begin{pmatrix} 0 & 1 & 0 & 0 & 0 & 0 \\ -1 & 0 & 0 & 0 & 0 & 0 \\ 0 & 0 & 0 & 1 & 0 & 0 \\ 0 & 0 & -1 & 0 & 0 & 0 \\ 0 & 0 & 0 & 0 & 0 & 1 \\ 0 & 0 & 0 & 0 & -1 & 0 \end{pmatrix} \begin{pmatrix} \partial H/\partial x \\ \partial H/\partial p_x \\ \partial H/\partial y \\ \partial H/\partial p_y \\ \partial H/\partial z \\ \partial H/\partial p_z \end{pmatrix} \quad (10)$$

where H is the Hamiltonian. From a solution of these equations of motion we know the relation between x, p_x, y, p_y, z, p_z at times T_1 and T_2:

$$\begin{pmatrix} x_2 \\ p_{x2} \\ y_2 \\ p_{y2} \\ z_2 \\ p_{z2} \end{pmatrix} = \begin{pmatrix} \frac{\partial x_2}{\partial x_1} & \frac{\partial x_2}{\partial p_{x1}} & \frac{\partial x_2}{\partial y_1} & \frac{\partial x_2}{\partial p_{y1}} & \frac{\partial x_2}{\partial z_1} & \frac{\partial x_2}{\partial p_{z1}} \\ \frac{\partial p_{x2}}{\partial x_1} & \frac{\partial x_2}{\partial p_{x1}} & \frac{\partial x_2}{\partial y_1} & \frac{\partial x_2}{\partial p_{y1}} & \frac{\partial x_2}{\partial z_1} & \frac{\partial x_2}{\partial p_{z1}} \\ \frac{\partial y_2}{\partial x_1} & \frac{\partial x_2}{\partial p_{x1}} & \frac{\partial x_2}{\partial y_1} & \frac{\partial x_2}{\partial p_{y1}} & \frac{\partial x_2}{\partial z_1} & \frac{\partial x_2}{\partial p_{z1}} \\ \frac{\partial p_{y2}}{\partial x_1} & \frac{\partial x_2}{\partial p_{x1}} & \frac{\partial x_2}{\partial y_1} & \frac{\partial x_2}{\partial p_{y1}} & \frac{\partial x_2}{\partial z_1} & \frac{\partial x_2}{\partial p_{z1}} \\ \frac{\partial z_2}{\partial x_1} & \frac{\partial x_2}{\partial p_{x1}} & \frac{\partial x_2}{\partial y_1} & \frac{\partial x_2}{\partial p_{y1}} & \frac{\partial x_2}{\partial z_1} & \frac{\partial x_2}{\partial p_{z1}} \\ \frac{\partial p_{z2}}{\partial x_1} & \frac{\partial x_2}{\partial p_{x1}} & \frac{\partial x_2}{\partial y_1} & \frac{\partial x_2}{\partial p_{y1}} & \frac{\partial x_2}{\partial z_1} & \frac{\partial x_2}{\partial p_{z1}} \end{pmatrix} \begin{pmatrix} x_1 \\ p_{x1} \\ y_1 \\ p_{y1} \\ z_1 \\ p_{z1} \end{pmatrix} \quad (11)$$

Inspecting the matrices J of Eq. 10 and the matrix A of Eq. 11 one finds by not too difficult transformations the so-called symplectic condition

$$J = AJA^t \quad (12)$$

to which Liouville's theorem is a subset. Writing the matrix A as function of the aberration coefficients of Table I, one finds many relations between those elements [13]. In the case of Table I as an example, all elements marked by stars can be expressed by the unmarked elements. Consequently [13] all but one of the flight-time-aberrations of any order are functions of lateral aberrations of equal or lower order. If all time-aberrations must vanish thus the corresponding lateral aberrations must vanish also. This implies that laterally focusing high resolving time-of-flight mass spectrometers must also be good imaging systems. It does not require the object and image to be small, but it requires that all trajectories starting from one point of the object are focused to one corresponding point of the image. An example in which this condition is fulfilled approximately is the system [7] shown in Fig.4, the aberration coefficients of which are listed in Table I. Since Eqs. 10,11 are applicable to any optical design, the given statement applies to sector field systems (see Figs. 1,3,4) as well as to clystron-like ion reflectors (see Fig.2).

Table I

$(x,x) = 1.000$	$(a,x) = 0.000$	$*(t,x) = -0.000$
$(x,a) = 0.000$	$(a,a) = 0.000$	$*(t,a) = -0.000$
$(x,\Delta) = 0.000$	$(a,\Delta) = 0.000$	$(t,\Delta) = 0.000$
$(x,xx) = 0.000$	$(a,xx) = 0.000$	$*(t,xx) = -0.000$
$*(x,xa) = -0.000$	$*(a,xa) = 0.000$	$*(t,xa) = -0.000$
$(x,aa) = 0.000$	$(a,aa) = -0.000$	$*(t,aa) = 0.000$
$(x,x\Delta) = 0.000$	$(a,x\Delta) = 0.004$	$*(t,x\Delta) = -0.001$
$(x,a\Delta) = -0.010$	$(a,a\Delta) = 0.000$	$*(t,a\Delta) = 0.000$
$(x,\Delta\Delta) = -0.000$	$(a,\Delta\Delta) = -0.004$	$(t,\Delta\Delta) = 0.002$
$(x,yy) = 0.000$	$(a,yy) = 0.000$	$*(t,yy) = 0.000$
$(x,yb) = 0.000$	$(a,yb) = -0.000$	$*(t,yb) = 0.000$
$(x,bb) = 0.000$	$(a,bb) = -0.000$	$*(t,bb) = 0.000$

For the time-of-flight mass spectrometer of Fig.4 the second order image aberrations are listed. In this design all lateral aberrations (x,...) as well as all longitudinal aberrations (t,...) are very small. As outlined in section 4 the aberration coefficients marked by "$_*$" are functions of the unmarked coefficients of equal or lower order. Note that only one longitudinal aberration coefficient of each order must be determined independently, i.e. (t,Δ), (t,$\Delta\Delta$), (t,$\Delta\Delta\Delta$) ...

References

1) B.A. Mamyrin, V.J. Karataev, D.V. Schmikk and V.A. Zagulin; Sov. Phys. JETP 37 (1973) 45
2) F.W. Aston; Phil. Mag. 38 (1919) 709
3) W. Bartky and A.J. Dempster; Phys. Rev. 33 (1929) 1019
4) H. Wollnik; Nucl. Instr. and Meth. 95 (1971) 453
5) W. Poschenrieder; Int. J. Mass Spectr. Ion Phys. 6 (1971) 413; 9 (1972) 39
6) H. Wollnik and T. Matsuo; Int. J. Mass Spectr. Ion Phys. 37 (1981) 209
7) J.M. Wouters, D.J. Vieira, H. Wollnik, H.A. Enge, S. Kowalski and K.L. Brown; Nucl. Instr. and Meth. (1985) in print
8) J. Mattauch und R. Herzog; Z. Phys. 89 (1934) 786; Phys. Rev. 50 (1936) 617
9) A.O. Nier; Rev. Sci. Instr. 18 (1947) 398
10) R.J. Conzemius and J.M. Capellen; Int. J. Mass Spectr. Ion Phys. 34 (1980) 197
11) A. Benninghoven ed.; Springer Ser. Chem. Phys., 25 (1983)
12) P.K.D. Feigl, F.R. Krueger and B. Schueler; Org. Mass Spectr. 18 (1983) 442
13) H. Wollnik and M. Berz; Nucl. Instr. and Meth. 238 (1985) 127

Design of an Organic SIMS Instrument with Separate Triple Stage Quadrupole (TSQ) and Time-of-Flight (TOF) Spectrometers

B.L. Bentz and R.E. Honig

RCA Laboratories, Princeton, NJ 08540, USA

1. Introduction

The need to characterize organic solids, e.g., polymers, has increased dramatically in recent years. That large organic molecules can be desorbed intact by various means and identified by their mass spectra was demonstrated by a number of workers during the past decade by bombardment with: keV particles [1-3]; MeV particles [4-6]; or photons [7,8]. A comparison of these methods shows that the resulting mass spectra are remarkably similar, as discussed more fully in a recent review paper [9]. The mass spectrometers employed in these studies were either quadrupole or time-of-flight instruments, matching in each case the characteristics of the primary particles employed.

If it is desired to identify organic molecules adsorbed on a substrate, the "Static SIMS" approach, as defined by BENNINGHOVEN [1] for surface analysis, appears most suitable. This method has been applied recently to various organic systems in a number of laboratories, e.g., BRIGGS and HEARN [10], CAMPANA et al. [11], BENNINGHOVEN [12], and GARDELLA and HERCULES [13]. This method limits the total primary ion dose to about 10^{13} particles/cm^2 so that less than 10 percent of the sample surface is damaged [14], thereby making sure that most molecules are not fragmented by multiple primary ion impact and thus can be desorbed intact from the surface. To compensate for this dose limitation, relatively large sample areas (~1 cm^2) are desirable. To identify the desorbed molecules and at the same time determine their structure, it is a logical step to combine static SIMS with a Triple Stage Quadrupole (TSQ) mass spectrometer. In such a system, the first quadrupole selects ions of a given mass; the second quadrupole serves as a reaction chamber where the selected ions are collisionally dissociated, and the third quadrupole scans the fragmentation spectrum, from which the molecular structure can be deduced.

Recently, several laboratories have combined dynamic SIMS with tandem mass spectrometers for organic analysis from liquid surfaces [15,16] and for characterizing polyatomic ions sputtered from inorganic solid surfaces [17,18]. We are unaware of any static SIMS/TSQ studies of organic solid systems.

The SIMS/TSQ combination will be investigated in detail with the instrument under construction, but there are two major drawbacks: the molecular mass range in our instrument is limited to 1000 daltons, and the instrumental sensitivity falls off substantially at higher masses which are of special interest here. For these reasons, the instrument has been designed so that it can also be used with a time-of-flight (TOF) mass spectrometer. The TOFMS has high, uniform transmission over its entire mass range, which is limited only by the efficiency of the ion detector for large masses. The major drawback of the TOF system is its low duty cycle for ion generation and detection. However, this will be off-set by the fact that we plan to detect secondary neutrals post-ionized by non-resonant multiphotons, as will be discussed in more detail below.

2. Design

2.1 General Considerations

The characterization of a small number of large organic molecules residing on a sample surface places stringent requirements on the overall efficiency of secondary ion production and subsequent mass analysis and detection. In this section, we shall compare the performance of a TSQ system with that of a TOF system.

The triple quadrupole approach utilizes a d.c. primary beam to produce a current of secondary ions which constitute a small fraction F_I of the total number of molecules desorbed from the sample. We shall assume $F_I \equiv$ secondary ions emitted/desorbed molecules = 10^{-4}. A fraction $F_A = 10^{-1}$ of the emitted ions passes the electrostatic energy filter and enters the first quadrupole Q1. Since usually Q1 is scanned over the entire mass spectrum (typically 10-1000 u), a given mass passes Q1 only a small fraction F_U of the time during a single scan, which results in $F_U \equiv$ ions utilized/ions transmitted = 10^{-3}. Furthermore, a quadrupole MS has a limited high mass transmission F_T, assumed here to be = 10^{-3}. This results in an estimated order-of-magnitude overall efficiency for the analysis of high masses by the triple quadrupole system

$$\eta_{TSQ} \equiv \text{ions detected/molecules desorbed}$$
$$= F_I \times F_A \times F_U \times F_T$$
$$= 10^{-4} \times 10^{-1} \times 10^{-3} \times 10^{-3} = 10^{-11}.$$

In our alternative approach, we utilize the desorbed neutral molecules, which constitute most of the sputtered particles, and post-ionize them. This is accomplished most effectively by non-resonant multiphoton ionization, using focused KrF or ArF laser pulses (photon energies: 5.0 or 6.5 eV; duration: $\cong$ 15 ns). Ionization efficiencies of 10% or better ($F_I > 10^{-1}$) can be expected [19]. For optimum sample utilization, the primary beam is pulsed at the pulse repetition frequency of the laser (10 Hz). The spectrometer ideally suited to match short laser pulses is the TOF mass analyzer. It has a high transmittance ($F_T \cong 10^{-1}$) independent of mass, and detects and records all masses simultaneously. The major disadvantage of this system is its low laser "duty cycle" ($F_D \equiv$ pulse length x repetition rate = $10^{-8} \times 10^{1} = 10^{-7}$). Thus for the same primary ion beam current as in the triple quadrupole approach, the overall efficiency for the laser post-ionization/TOF system is given by

$$\eta_{TOF} \equiv \text{ions detected/molecules desorbed}$$
$$= F_I \times F_D \times F_T = 10^{-1} \times 10^{-7} \times 10^{-1} = 10^{-9}.$$

The order-of-magnitude estimates presented above suggest that a laser post-ionization/TOF combination should be well suited to analyze extremely small samples. The TOFMS has an unlimited mass range and two additional advantages if constructed as a "Reflectron" [20]: its resolving power ($M/\Delta M$) can be on the order of several thousand, and metastable molecules can be detected at its mid-point. On the other hand, the TSQ system will provide more detailed structural information, but its mass range is limited, and its high mass sensitivity is poor. For these reasons, the proposed design provides for a dual approach, which includes both a Finnigan-MAT triple quadrupole mass spectrometer and a time-of-flight analyzer to be used separately, not in tandem or simultaneously. The essential system components are shown, drawn to scale, in Fig. 1.

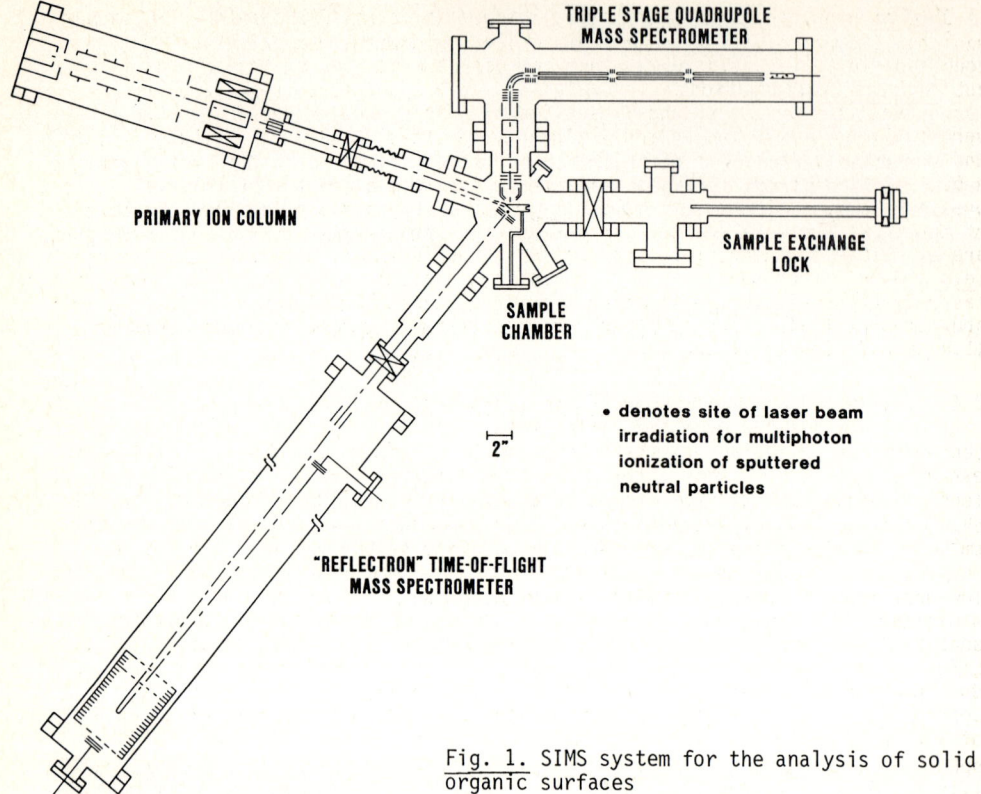

Fig. 1. SIMS system for the analysis of solid organic surfaces

2.2 Primary Ion Column

The primary ion source used to produce keV ions is a "Colutron G-2" gas discharge source equipped with an ExB mass filter for primary beam selection. Argon or xenon serve as discharge gases to form primary ion beams with energies up to 10 keV. The source is pumped by a 500 l/sec turbopump, while a small ion pump, situated on the high vacuum side of the source aperture, further reduces gas load in the column. In order to reduce the background pressure in the analysis chamber, the source can be modified for cesium ion production [21]. A $2°$ bend has been inserted in the beam line to prevent high-energy neutral particles, produced by charge exchange, from reaching the target. All apertures in the column are electrically isolated to allow measurement and optimization of beam currents. A condenser and an objective lens control the beam current delivered to the sample and the final beam size at the sample. The final 3-element objective lens is of asymmetric geometry to aid the formation of small (<30 μm) beam probes. The fluence condition for static SIMS can be met by either forming a defocused static beam or scanning a focused probe over a relatively large sample area. Raster scanning and beam deflection are performed by plate condensers placed between the objective lens and sample. The ion beam operates in static or rastered, d.c. or pulsed modes. Primary ions strike the target at an angle of $70°$ with respect to the target normal.

2.3 SIMS Analysis Chamber

A 15-port stainless steel chamber, designed and constructed in-house, serves as the SIMS UHV vessel. The cylindrical chamber has a nominal o.d. of 15.2 cm and is 35.6 cm high. As shown in Fig. 1, it couples to the Finnigan-MAT TSQ vacuum manifold (pumped by 500 l/sec and 300 l/sec turbopumps) via a 15.2 cm o.d. double-sided Conflat flange machined to support a portion of the secondary ion optics. The SIMS chamber is evacuated by a 500 l/sec turbopump fitted with a large area liquid N_2 trap and Ti sublimator. This pumping stack is mounted vertically on top of the cylindrical chamber. Metal seals are used throughout, and the chamber is bakeable to 200 C. A homemade sample transfer lock, pumped by a 50 l/sec turbopump, allows samples to be inserted into the main chamber without breaking vacuum. The sample holder is electrically isolated to allow measurement of target currents. The ports on the chamber are arranged on three levels; ports are available for electrical feedthroughs, supplementary particle beam sources (e.g., an electron gun), viewing optics, and windows for the entry and exit of a laser beam for either sample desorption or gas-phase multiphoton ionization of sputtered neutrals. The TOF tube communicates with the main chamber through a 11.4 cm o.d. Conflat flange.

2.4 Secondary Ion Transfer and TSQ Mass Analysis

Secondary ions are accelerated from the sample surface by an extraction field established by an immersion lens positioned close to the sample [22,23]. The lens is designed to collect secondary ions with initial kinetic energies up to 10 eV_2 emitted in a 15^0 angular half-cone. The sampling area is approximately one cm^2. A suitable geometry and operating potentials to achieve this end are modeled by computer, using an electron optics program developed at RCA Labs. For low-dose static SIMS applications where beam damage to the sample is to be minimized, it is desirable to sputter a large area of the sample to enhance sensitivity. To this end, dynamic emittance matching is used, as described by LIEBL [24]. In this scheme, the secondary ion beam transfer optics includes a double deflector system synchronized with the primary beam raster. This method continuously directs secondary ions onto their optical axis, which has the effect of increasing the acceptance angle of the mass spectrometer [25,26]. A 90^0 spherical electrostatic analyzer (mean radius: 30.5 mm; energy window $\Delta E/E$ = 4%) filters out high-energy neutral particles and photons from the secondary ion beam. Deceleration and focus lenses preceed and follow this device to adjust beam energies for the spherical sector and for the first quadrupole of the triple quadrupole system, respectively. The circular rods comprising the triple quadrupole stack are each 122 mm long and 6.3 mm in diameter. The mass range of the instrument is 4-1000 u. The mass-analyzed ions are detected by a high-gain channel electron multiplier operated in the pulse counting mode. The axis potential of the entire system is shown in Fig. 2.

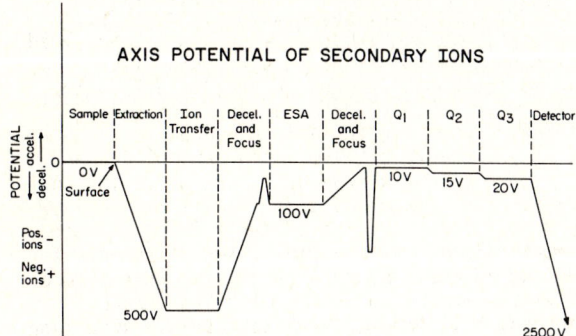

Fig. 2. Axis potential in SIMS/TSQ system

2.5 Excimer Laser and Time-of-Flight Spectrometer

To post-ionize the sputtered neutrals that are to be analyzed in the TOFMS, a UV laser, such as an excimer laser filled with either KrF (λ = 248 nm; E = 5.0 eV) or ArF (λ = 193 nm; E = 6.5 eV) is required. The former makes it simpler to

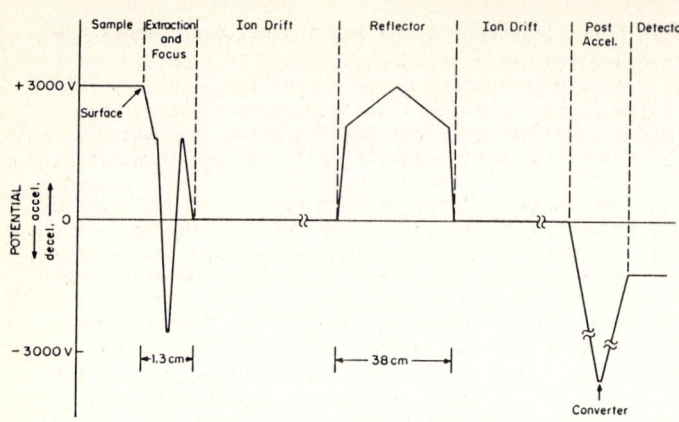

Fig. 3. Axis potential in SIMS/TOF system

AXIS POTENTIAL OF SECONDARY IONS

guide the laser beam without losses outside the vacuum chamber; the latter with its larger photon energy ionizes more effectively. The dot shown in Fig. 1 below the tilted sample position indicates the location of the laser beam normal to the paper plane. Preliminary tests have shown that an existing TACHISTO laser (1979 vintage) is fully operative and adaptable to our present requirements. However, it is expected that eventually a state-of-the-art laser of adequate power and higher repetition rate should be employed, in order to increase detection sensitivity.

The TOF spectrometer is a reflector-type instrument having a total flight path of 3.2 meters. The system will detect either secondary <u>ions</u> desorbed from the bombarded sample surface, or sputtered <u>neutral</u> particles which will be ionized in the gas phase by a focused UV laser beam. Fig. 3 shows the TOF axis potential. The pulsed primary beam produces ion packets, tens of nanoseconds to microseconds long, which bombard the sample mounted on a dual sample holder held at several kilovolts above ground (see Fig. 1). The extraction field for ion collection is set up by a single, high transmission grid at ground potential, although we plan to experiment with a multielement accelerating lens which can provide some collimation of the secondary ion bundle. The ion beam, deflected approximately 2°, enters a Mamyrin-type two-stage electrostatic mirror [20]. Use of a mirror enhances mass resolution by providing first- and second-order energy focusing. Chevron-type multichannel plates are mounted on axis behind the mirror. This detector assists optimum tuning of the extraction and transport optics, and, when used together with the reflected ion detector, can serve to study the decomposition of metastable species [27]. Ions reflected by the mirror are post-accelerated by an additional 5 keV and detected by a multichannel plate detector operated in the pulse-counting mode.

References

1. A. Benninghoven, D. Jaspers, and W. Sichtermann: Appl. Phys. <u>11</u>, 35 (1976)
2. M. Barber, R. S. Bordoli, R. D. Sedgwick, and A. N. Tyler: J.C.S. Chem. Comm. 325 (1981)
3. M. Barber, R. S. Bordoli, R. D. Sedgwick, and A. N. Tyler: Nature <u>293</u>, 270 (1981)
4. R. D. Macfarlane and D. F. Torgerson: Science <u>191</u>, 920 (1976)
5. P. Dück, W. Treu, W. Galster, H. Fröhlich, and H. Voit: Nucl. Instr. Methods <u>168</u>, 601 (1980)
6. P. Håkansson, A. Johansson, I. Kamensky, B. Sundqvist, J. Fohlman, and P. Peterson: IEEE Trans. Nucl. Sci. <u>NS-28</u>, 1776 (1981)
7. E. Unsöld, F. Hillenkamp, and R. Nitsche: Analusis <u>4</u>, 115 (1976)

8. M. A. Posthumus, P. G. Kistemaker, H. L. C. Meuzelaar, and M. C. Ten Noever de Brauw: Anal. Chem. 50, 985 (1978)
9. R. E. Honig: Int. J. Mass Spectrom. Ion Proc. 66, 31 (1985)
10. D. Briggs and M. J. Hearn: Spectrochim. Acta 40B, 707 (1985)
11. J. E. Campana, M. M. Ross, S. L. Rose, J. R. Wyatt, and R. J. Colton: in "Ion Formation from Organic Solids", A. Benninghoven, Ed. (Springer Verlag, Berlin, 1983), p. 144
12. A. Benninghoven: J. Vac. Sci. Technol. A3, 451 (1985)
13. J. A. Gardella, Jr. and D. M. Hercules: Anal. Chem. 52, 226 (1980)
14. C. W. Magee: Int. J. Mass Spectrom. Ion Phys. 49, 211 (1983)
15. D. F. Hunt, W. M. Bone, J. Shabanowitz, J. Rhodes, and J. M. Ballard: Anal. Chem. 53, 1704 (1981)
16. G. L. Glish, P. J. Todd, K. L. Busch, and R. G. Cooks: Int. J. Mass Spectrom. Ion Proc. 56, 177 (1984)
17. C. W. Magee, P. J. Gale, B. L. Bentz, and J. Shabanowitz, presented at the 30th Annual Conference on Mass Spectrometry and Allied Topics, Honolulu, Hawaii, 1982, p. 523
18. T. F. Magnera, D. E. David, R. Tian, D. Stulik, and J. Michl, J. Am. Chem. Soc. 106, 5040 (1984)
19. C. H. Becker and K. T. Gillen: J. Opt. Soc. Am. B 2, 1438 (1985)
20. B. A. Mamyrin, V. I. Karataev, D. V. Shmikk, and V. A. Zagulin: Sov. Phys.-JETP 37, 45 (1973)
21. C. W. Magee: J. Electrochem. Soc. 126, 660 (1979)
22. C. W. Magee, W. L. Harrington, and R. E. Honig: Rev. Sci. Instrum. 49, 477 (1978)
23. K. J. Hanszen and R. Lauer: in "Focusing of Charged Particles", A. Septier, Ed. (Academic, New York 1978), p. 296
24. H. Liebl: in Inst. Phys. Conf. Ser. No. 38, "Low-Energy Ion Beams", K. G. Stephens, I. H. Wilson, and J. L. Maruzzi, Eds. (Institute of Physics, London 1978), p. 266
25. H. Liebl: Nucl. Instrum. Methods 187, 143 (1981)
26. J. E. Campana, J. J. DeCorpo, and J. W. Wyatt: Rev. Sci. Instrum. 52, 1517 (1981)
27. S. Della-Negra and Y. Le Beyec: Anal. Chem. 57, 2035 (1985)

High-Resolution TOF Secondary Ion Mass Spectrometer

E. Niehuis, T. Heller, H. Feld, and A. Benninghoven

Universität Münster, Physikalisches Institut, Domagkstr. 75,
D-4400 Münster, F. R. G.

1. Introduction

The development of soft ionization methods extended mass analysis to large involatile biomolecules. In many applications, especially in life sciences, only very limited amounts of sample material are available. Time-of-flight mass analyzers meet the requirement of an extremely high sensitivity due to their high transmission and quasi-simultaneous detection of all masses. Detection limits in the pico- and femtomol range can be achieved by TOF-SIMS for a variety of organic samples /1/.

Another advantage of the TOF method is the unlimited mass range. However, for high-mass applications the low-mass resolution of most instruments is a severe limitation. For a reflectron type TOF analyzer a mass resolution above 10,000 can be expected. The required overall time resolution is directly related to the size of such an instrument. In addition, it determines the time of analysis in the ion counting mode via the repetition rate. Fast timing electronics with 1 ns time resolution are available and therefore a primary pulse width of about 1 ns is desirable. A mass separation in the primary system is necessary in order to get a high dynamic range. Any contamination in the primary beam will distort the secondary ion spectrum due to flight time differences in the primary system.

This paper presents some of the first results obtained with a new high-resolution TOF-SIMS instrument equipped with a mass separated pulsed primary ion source and a reflectron TOF analyzer.

2. Design of the instrument

2.1. Primary beam system

We developed a new electrodynamic mass separation and beam chopping method based on a pulsed 90 degree deflection. The same momentum, defined by the time integral of the HV deflection pulse, is added to primary ions of different mass but same energy. Therefore their final momenta have different directions. The diverging angle is drastically reduced after the 90 degree deflection, because radial velocity differences are transformed into axial ones. This provides the separation of different masses by an appropiate drift path. The mass resolution can be as high as 1000.

An electron impact ion source is used to produce a continuous beam of noble gas ions with 10 keV energy. The deflection field is switched on for 1 µs by a transistorized pulse generator with 6 kV amplitude and 20 ns rise- and falltime. Before the ions leave the deflector the pulse is switched off. At the exit slit of the deflector a bunch of ions, perpendicular to the direction of motion, is cut out of the beam. A drift path of 25 cm with an exit slit of 1 mm is used for the mass separation. The beam is focused onto the target by an Einzel-lens with a spot diameter of 50 µm.

The diverging angle is reduced at the cost of an increase in axial velocity differences. Hence the ion bunch spreads, resulting in a pulse width of about 20 ns at the exit of the deflector. These velocity differences can be reversed by axial bunching. A second HV pulse is switched on when the ion bunch is in the center of the bunching system. The ions gain energy according to their position. Equal flight times to the target can be achieved by adapting the amplitude of the bunching pulse. The pulse width at the target depends on the entrance slit width of the deflector and the diverging angle of the incident ion beam. A pulse width of 1.3 ns is expected for 0.5 mm and 0.5 degrees. An increase to 1.5 ns is caused by the nonnormal incidence on the target.

2.2 Mass analyzer

High-resolution time-of-flight SIMS requires energy isochronous focusing. Several types of electrostatic TOF-MS have been proposed using mirrors and sector fields. Second order energy focusing is achieved by a two-stage reflector /2/. Multiple time and space focusing sector field instruments have been designed by POSCHEN-RIEDER /3/ and MATSUDA /4/. As the mass resolution of the proposed stigmatic imaging sector field instruments is limited by the quadratic errors of the entrance angles, we have chosen the reflectron-type analyzer.

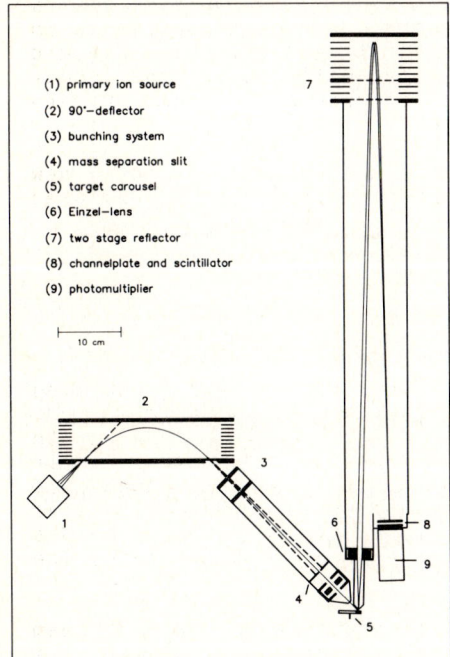

The reflection angle of the analyzer is 177 degrees; the mirror consists of a retarding gap (30 mm) and a reflecting gap (60 mm) defined by grids of 90 % transparency. The total effective flight path is 2.2 m. In order to improve the transmission an Einzel-lens is used for the beam transport. This lens provides a large acceptance angle of +/- 1.5 degrees in both directions at a magnification of 50.

The conditions for second order energy focusing are described in the literature /2/ but in a SIMS reflectron the additional time dispersion in the acceleration gap limits the mass resolution. For a restricted energy range this time dispersion can be reduced by tuning the reflecting field to a lower field strength compared to second order energy focusing conditions. A mass resolution of about 15,000 can be expected for an acceleration gap of 2 mm at an energy width of 0.4 %.

Fig. 1 Schematic of the instrument

2.3 Detection system

The detection system is similar to the one used in our first TOF SIMS /5/. It provides a postacceleration of the secondary ions up to 20 keV energy for both polarities in order to achieve a high detection efficiency even for large organic molecules. A single channel plate is used for the ion-electron conversion and preamplification. The signal is coupled to ground and amplified by a fast scintillator and a photomultiplier. The photomultiplier is mounted outside the vacuum chamber and its exit is at ground potential.

A new registration system with a time resolution of 1.25 ns is under construction. It is based on a triggerable time-to-digital converter with 256 stops and a special hardware for the fast accumulation of the data. Count rates of more than 1E6 c/s can be processed.

3. First results

3.1. Primary beam system

The mass separation of the primary system is demonstrated by a Xe^{++} spectrum (fig. 2). The primary ion spectrum has been recorded by sweeping the amplitude of the HV pulse. The mass resolution of 500 achieved with a drift path of 70 cm and an exit slit width of 1 mm is in good agreement with the calculations.

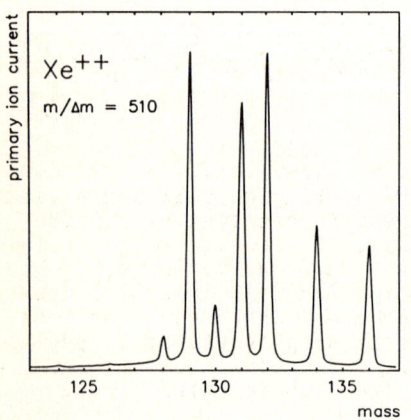

Fig.2
primary ion mass spectrum

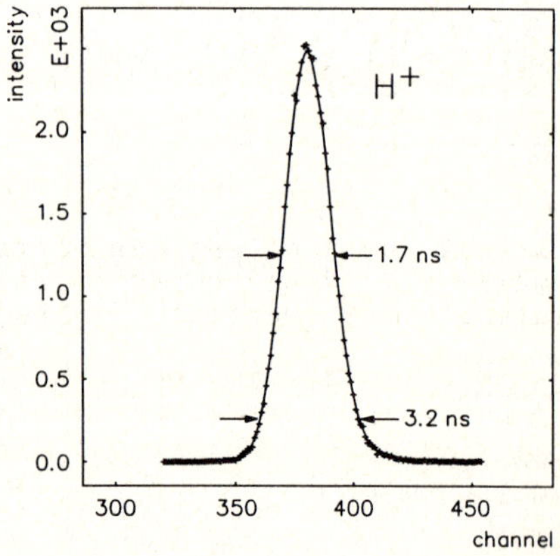

Fig.3
H^+-peak of the SI mass spectrum

The peak width of the H^+-peak in the secondary ion spectrum reflects the primary pulse width (fig. 3). It has been measured with a time-to-pulse-height converter (TAC) combined with a multichannel analyzer (MCA). A 100 MHz clock synchronized to the primary pulse was used to create the delayed start of the TAC. The peak width of 1.7 ns includes the time dispersion in the analyzer. Taking this into account, the resulting primary pulse width is 1.5 ns.

3.2 Mass analyzer

The resolution was tested with the TAC-MCA. The base peak of the Crystal Violet spectrum (M+=372.24) has a total flight time of

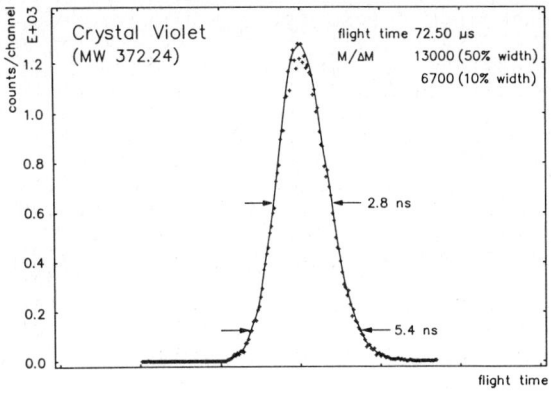

Fig. 4 Base peak of Crystal Violet prepared on etched silver

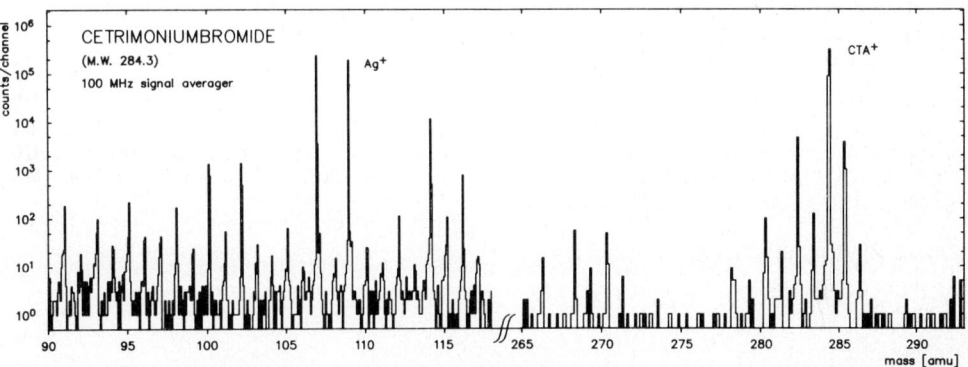

Fig. 5 Pos. SI spectrum of Cetrimoniumbromide, prepared on etched silver

72.50 µs at an acceleration voltage of 2.0 kV. The peak width at half maximum (2.8 ns) corresponds to a mass resolution of 13,000, the width at tenth maximum (5.4 ns) to 6700.

The dynamic range is demonstrated by the logarithmic spectrum of Cetrimoniumbromide (fig. 5). This spectrum has been recorded with the triggerable 100 MHz signal averager of our first TOF-SIMS. A dynamic range of six orders of magnitude can be achieved due to the mass separation of the primary system. The quality of the primary pulse and of the energy focusing analyzer is reflected by the peak shape of the Ag 108.9 peak; the peak width 3.5 orders of magnitude below the maximum is smaller than 10 ns.

4. Conclusion

A pulsed 90°-deflection was presented for beam chopping and primary ion mass separation. This method can be applied to all types of ion sources. Due to mass separation the primary pulse is well defined providing a high dynamic range of 6 orders of magnitude in the secondary ion spectrum. The pulse width of 1.5 ns is matched to the time resolution of modern fast time-to-digital converters.

A mass resolution of 13,000 was achieved by a reflectron type TOF-SIMS instrument which is yet rather compact (overall length of the analyzer 1 m). This is the highest mass resolution reported for an electrostatic TOF-MS. Due to the high transmission and quasisimultaneous detection of all masses the sensitivity of the instrument is extremely high. This is most important for applications where only very limited amounts of sample material are available, for example static SIMS applications in organic trace analysis.

References

1. A. Benninghoven, E. Niehuis, T. Friese, D.Greifendorf
   and P.Steffens: Org. Mass Spectrometry 19 (1984) 346
2. V.I. Karataev, B.A. Mamyrin and D.V. Shmikk;
   Sov. Phys.-Tech. Phys., 16 (1972) 1177
3. W.P. Poschenrieder
   Int. J. Mass Spec. Ion Phys.,9 (1972) 357
4. T. Sakurai, T. Matsuo and H. Matsuda
   Int. J. Mass Spec. Ion Phys., 63 (1985) 273
5. P. Steffens, E. Niehuis, T. Friese, D. Greifendorf and
   A. Benninghoven: J. Vac. Sci. Technol. A 3 (1985) 1322

Part VII

Fourier Transform Ion Cyclotron Resonance

Laser Desorption Fourier Transform Mass Spectrometry: Mechanisms of Desorption and Analytical Applications

M.P. Chiarelli, D.A. McCrery, and M.L. Gross
Department of Chemistry, University of Nebraska, Lincoln, NE 68588, USA

1. Introduction

Mechanistic aspects of laser desorption (LD) polar organics are the subject of intense investigation by many research groups [1,2]. There exist two broad classes of experimental conditions for LDMS [3]. One involves low power, broad beam lasers with thick (>1 um) sample layers. The other makes use of a highly focused laser (1 um^2) and thin sample films. Similarly, there are two broad categories of desorption mechanism: thermal and non-thermal. Low power densities, long pulse lengths, and thick sample layers are conditions that favor thermal desorption; i.e., a process occurring under the influence of sample temperature.

Both Joule heating and LD lead to thermal desorption of intact quaternary ammonium salts [4-6]. Adduct ions of K^+ and sucrose also have been seen under LD and Joule heating conditions [7]. The kinetic energy distributions of sodium sucrose ions generated from a moderately focused laser correspond to temperatures three times that generated in the substrate by the laser shot. A probable mechanism of increasing the kinetic energy of the selvedge-formed ions was cited as being due to a rapid loss of electrons from a "cold plasma" above the substrate, leading to positive ion acceleration [8].

Many different processes could be described as non-thermal LD. SEYDEL and LINDER [2,9] proposed, for conditions normally employed with LAMMA, that desorption is initiated by a laser driven shock wave. HERCULES [10] postulated that several processes occur, their relative importance dictated by the distance from the laser shot. HILLENKAMP [11] proposed that "collective, non-equilibrium processes in the condensed phase" were responsible for desorption.
Gas-phase complexation has strong support as a dominant mechanism for ion formation under thermal conditions. VAN DER PEYL et al. [7] showed that the gas-phase complexation of K^+ and sucrose is a feasible process. STOLL and ROLLGEN [12] employed a low power laser focussed on two spatially separated substrates, and showed support for a gas-phase complexation mechanism. As power density increases, however, the extent of gas-phase reaction becomes less certain. HARDIN and VESTAL [13] suggest that ion formation occurs from a breakdown of metastable clusters.

Similar aspects of all types of desorption spectra have prompted COOKS and BUSCH [14] to propose two totally separable desorption/ionization processes. Desorption is facilitated by the fast cleavage of intramolecular attractions in the solid, and complexation occurs in the selvedge.

In this report, we demonstrate the utility of FTMS for determination of several different classes of biomolecules which require metal ion attachment for ionization and discern mechanistic aspects of LD as they pertain to FTMS detection. It has been noted that changing experimental parameters in transmission instruments strongly influences the desorption process [8,11]. Therefore, investigations are needed to place LD-FTMS in the context of other mechanistic studies.

2. Experimental

The LD spectra were obtained with a home-built Fourier transform mass spectrometer interfaced to a Nicolet FTMS-1000 computer and data system as described previously [15,16]. Sample solutions, containing 1 mg/ml and equal weight of NaCl when appropriate, were either pipetted or electrosprayed on to a .33cm diameter probe tip as described elsewhere [17]. A Quanta-Ray DCR-2 Nd:YAG laser was used in both Q-switched and non-Q-switched modes, with pulse widths of 9 ns and 150 us, respectively, at a wavelength of 1064 nm.

3. Biomolecule Determination

Before discussing mechanism, it is useful to review a few applications in order to demonstrate some capabilities of LD-FTMS as to prepare a groundwork of experimental facts for the mechanism discussion.

3.1 Nucleosides

Nucleosides have been extensively studied with other forms of desorption ionization [18,19]. Although nucleosides may be desorbed as negative ions, a fruitful means of producing ions is by cationization.

Without addition of alkali metal salts, cationized nucleosides are desorbed, and this suggests that cationization occurs in the gas phase. A full report on the use of LD-FTMS for nucleosides, oligosaccharides, and glycosides will be published elsewhere [20].

3.2 Oligosaccharides and Glycosides

Although monosaccharides produce abundant $(M-H)^-$ ions and fragments resulting by losses of H_2O and CH_2OH, disaccharides do not yield a detectable $(M-H)^-$ under the conditions of LD-FTMS. Clearly cationization is required.

In the positive ion mode, disaccharides give $M+Na^+$, K^+, or Cu^+ adduct ions when a copper substrate is used. Melibose, an example of non-reducing sugars, undergoes fragmentation within the sugar moiety and yields ions due to losses of CH_2O groups. Sucrose, a reducing sugar, (see Fig. 1) shows primarily glycosidic cleavage. The molecular sodium complex is the most abundant ion if the sample is doped with NaCl. This is also true of the disaccharide lactose.

The success with disaccharides prompted the investigation of more complex glycosides such as digoxin, digitoxin, and erythromycin [20]. Samples were investigated in the non-Q-switched mode. The longer irradiation time produced salutary results for oligosaccharides as well as glycosides. The first two show cleavages of the saccharide chain, indicating that Na^+ is bound to the sugar portion of the molecule. The erythromycin-sodium complex (see Fig. 2) shows losses of the sugar cladinose, H_2O, and both. All three glycosides give the molecular adduct with Na^+ as the most abundant ion.

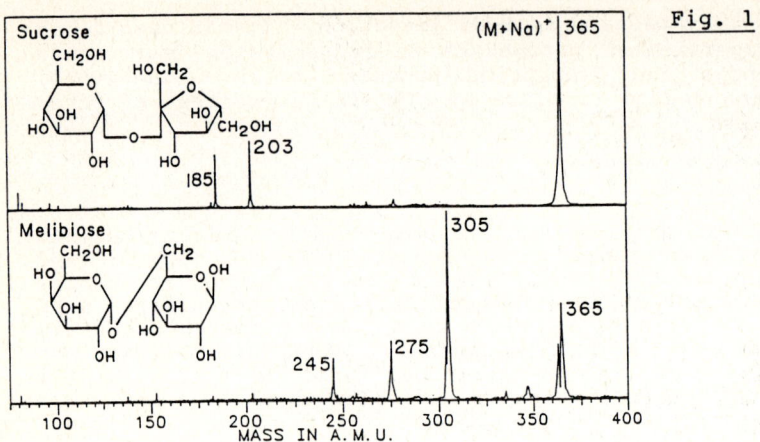

Fig. 1

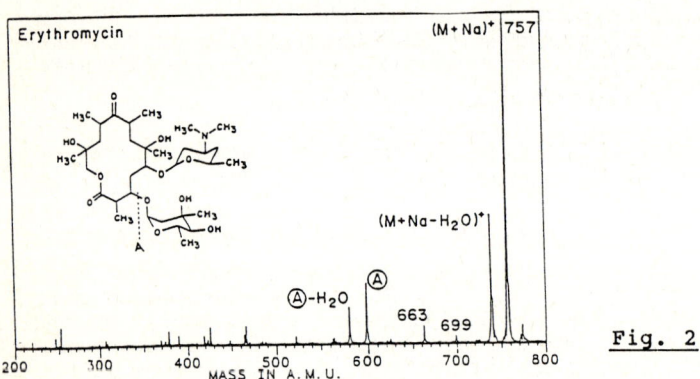

Fig. 2

4. Reproducibility and the Extent of Fragmentation

It has been noted that different sample preparations influence the reproducibility and extent of fragmentation seen in transmission instrument mass spectra [11]. We have investigated [21] the effect of sample preparation on the reproducibility and degree of fragmentation of N-propyltriphenylphosphonium bromide and sodium sucrose under four different conditions [21]. Samples were either pipetted or electrosprayed onto a smooth or rough probe surface.

The RSD values for ion relative abundances associated with the two methods of deposition are similar for the phosphonium salt (ca. 15% smooth, 22% rough, for the abundance ratio m/z 263/289) and slightly worse than the precision of EI spectra. The precipitation of the salt from solution forms uniform layers when pipetting and, of course, electrospray does not improve the homogeneity of the deposition for a material which already is ionic in the solid state.

The precision of the abundance ratio for $(M+Na)^+$ and the fragment m/z 203 produced by glycosidic cleavage could be improved by electrospraying the sample onto a smooth surface in lieu of pipetting. The electrospray improves the homogeneity of the deposition and the efficiency of the subsequent codesorption.

Differences in fragmentation result from desorption from a smooth vs. rough substrate. More fragmentation of the phosphonium ion occurred when the salt was desorbed from a rough surface. We suggest that heating rates are greater at the protrusions of the rough probe, and this leads to an increase in the ions' internal energy. However, the abundances of fragment ions in the mass spectrum of desorbed (sucrose + Na)$^+$ are less when a rough surface substrate was used. Higher localized protrusion temperatures allow for more sucrose to desorb, and volatilization becomes more competitive with fragmentation. The desorption of phosphonium ions is more nearly thermal, and occurs ar lower average power than does (sucrose + Na)$^+$.

All the results discussed in this section are suggestive of sucrose cationization in the solid state or in a plasma near the substrate surface, but not in the diffuse gas-phase. Diffuse gas-phase cationization can also be ruled out by suitable double resonance experiments. Ejection of Na$^+$ and Na$_2$Cl$^+$ had no effect on the abundance of (sucrose + Na)$^+$. Furthermore, if (sucrose + Na)$^+$ were ejected immediately after the laser beam, no new (sucrose + Na)$^+$ was formed at long times.

6. Split Probes

The absence of any diffuse gas-phase reactions prompted us to perform more spatially resolved experiments. In this sequence of experiments we utilize a split probe to ascertain adduct formation as a function of distance between the solid phases of NaCl and sucrose.

NaCl was sprayed on a bulk copper substrate and sucrose on a stainless steel mesh, and both substrates were placed coincident with the laser beam. The formation of m/z 365 was monitored at several different distances. The laser was operated in the Q-switched mode at an approximate power density of 10^8 W/cm^2. Control experiments were run to gauge the extent of attenuation caused by the mesh, i.e., NaCl/sucrose was desorbed off the copper with and without the mesh present and then desorbed off the mesh only.

It was found at a distance of 2.5 mm that there was no detectable (Na + sucrose)$^+$, m/z 365. At a distance of .1 mm, m/z 365 was 25% of the intensity of that measured in the control experiment (uncoated mesh present) run at that distance of separation. This observation is taken as support that the two selvedge regions are beginning to overlap. It should be noted that these are preliminary experiments, and a precise determination of the distance when overlap occurs remains to be determined. The necessary experiments are under way in our laboratory.

7. Acknowledgement

This research was supported by the U.S. National Science Foundation (Grant CHE-8018245) and the National Institutes of Health (Grant GM-30604).

8. References

1. G.J.Q. van der Peyl, J. Haverkamp, and P.G. Kistemaker: Int. J. Mass Spectrom. Ion Phys. 42 (1982) 125-141.
2. B. Linder and U. Seydel: Anal. Chem. 57 (1985) 895-899.

3. F. Hillenkamp: in Ion Formation from Organic Solids, as in (9).
4. R. Stoll and F.W. Rollgen: J.C.S. Chem. Commun. 16 (1980) 789.
5. R.J. Cotter and A.L. Yergey: J. Am. Chem. Soc., 103 (1981) 1596-1598.
6. R.B. Van Bremen, M. Snow, and R.J. Cotter: Int. J. Mass Spectrom. Ion Phys. 49 (1983) 35-50.
7. G.J.Q. van der Peyl, K. Isa, J. Haverkamp, and P.G. Kistemaker: Org. Mass Spectrom. 16 (1981) 416-420.
8. G.J.Q. van der Peyl, W.J. van der Zande, and P.G. Kistemaker: Int. J. Mass Spectrom. Ion Phys. 62 (1984) 51-71.
9. U. Seydel and B. Linder: in Ion Formation from Organic Solids, ed. by A. Benninghoven (Springer-Verlag, Berlin, 1983) 240.
10. D.M. Hercules, R.J. Day, K. Balasanmugam, T.A. Dang, and C.P. Li: Anal. Chem. 54 (1982) 280-305A.
11. F. Hillenkamp: Int. J. Mass Spectrom. Ion Phys. 45 (1982) 305-313.
12. R. Stoll and F.W. Rollgen: Z. Naturforsch. 37a (1982) 9-14.
13. E.D. Hardin and M.L. Vestal: Anal. Chem. 53 (1981) 1492-1497.
14. R.G. Cooks and K.L. Busch: Int. J. Mass Spectrom. Ion Phys. 53 (1983) 111-124.
15. E.B. Ledford, Jr., R.L. White, S. Ghaderi, C.L. Wilkins, and M.L. Gross: Anal. Chem. 52 (1980) 2450-2451.
16. D.A. McCrery, D.A. Peake, and M.L. Gross: Anal. Chem. 57 (1985) 1181-1186.
17. C.J. McNeal, R.D. Macfarlane, E.L. Thurston: Anal. Chem. 51 (1979) 2036-2039.
18. F.W. Crow, K.B. Tomer, M.L. Gross, J.A. McCloskey, and D.E. Bergstrom: Anal. Biochem. 139 (1984) 243-262.
19. C.J. McNeal, K.K. Ogilvie, N.Y. Theriault, M.J. Nemer: J. Am. Chem. Soc. 104 (1982) 976-980 and 981-984.
20. D.A. McCrery and M.L. Gross: Anal. Chim. Acta, in press.
21. D.A. McCrery and M.L. Gross: Anal. Chim. Acta, in press.

Desorption Ionization and Fourier Transform Mass Spectrometry for the Analysis of Large Biomolecules

D.H. Russell and M.E. Castro

Department of Chemistry, Texas A & M University,
College Station, TX 77843, USA

1. Introduction

The recent developments in biomolecule mass spectrometry can be traced to progress made in new ionization methods and instruments having extended mass range. Early in the development of Fourier transform mass spectrometry (FTMS) Comisarow and Marshall [1] Gross and Wilkins [2] and McIver [3] noted the applicability of this method for the analysis of large molecules. The principle limitation to developing these capabilities was in adapting the available desorption ionization methods to FTMS. High-mass FTMS requires the use of super-conducting magnets, the design of which imposes mechanical limitations in designing such systems, and ultra-high vacuum (less than 10^{-8} torr) which is not compatible with many desorption ionization sources and methods, e.g., gaseous discharge primary ion sources and liquid matrices. The first successful demonstration of desorption ionization with FTMS was the laser desorption ionization by Gross [4]. Subsequent to this work our laboratory reported on the use of Cs^+ ion SIMS with FTMS detection of high-mass ions, e.g., greater than m/z 2000 [5]. These preliminary results demonstrated the feasibility for performing desorption ionization with FTMS, and more recent work from these two laboratories clearly establishes the analytical utility of the method. Also, Wilkins [6] has reported impressive results for laser desorption-FTMS of a variety of biomolecules, and McIver and Hunt [7,8] have obtained excellent data on biomolecules with tandem quadrupole-FTMS system and fast-atom bombardment (FAB) ionization.

In this paper results of studies on Cs^+ ion desorption ionization-FTMS on biomolecules will be discussed. These studies illustrate the present capabilities of the method, but we will also point out problems which must be addressed to refine the instrumentation.

2. Results and Discussion

Our early work with Cs^+ ion DI-FTMS emphasized the nature of the secondary ions produced and the influence of the experimental parameters, e.g., beam density and irradiation time, on the yield of molecular, cluster and fragment ions. Detailed studies on the Cs^+ ion DI-FTMS spectra of sugars, viz. B-cyclodextrin, showed that formation of molecular ions, e.g., $[M+H]^+$, was favored at low beam density (less than 10^{-9} A/cm^2), and that cluster ions, e.g., $[M+Na]^+$ and/or $[M+xNa-(x-1)H]^+$, or fragments of cluster ions are formed at higher (>50-100 x 10^{-9} A/cm^2) beam densities [9]. Similar results have been obtained for small peptides. The ease with which organo-alkali metal cluster ions are formed with Cs^+ ion DI-FTMS has led us to investigate in some detail the chemistry of these species. An additional motivation for these studies was the observation that the addition of KCl to samples of small peptides produced abundant $[M+K]^+$ and $[M+xK-(x-1)H]^+$ ions, but that all the fragment ions in the spectrum are formed by fragmentation of the $[M+Na]^+$ ions. Based on information from a variety of studies, e.g., collision-induced dissociation, ion-molecule reaction chemistry, and etc., we attribute this to the relative stability of the ionic complex for $[M+Na]^+$ and $[M+K]^+$. That is, the binding energy of Na^+ is roughly twice that for

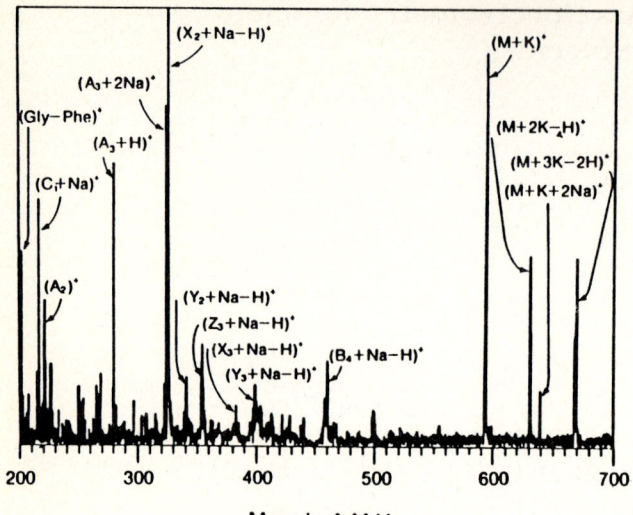

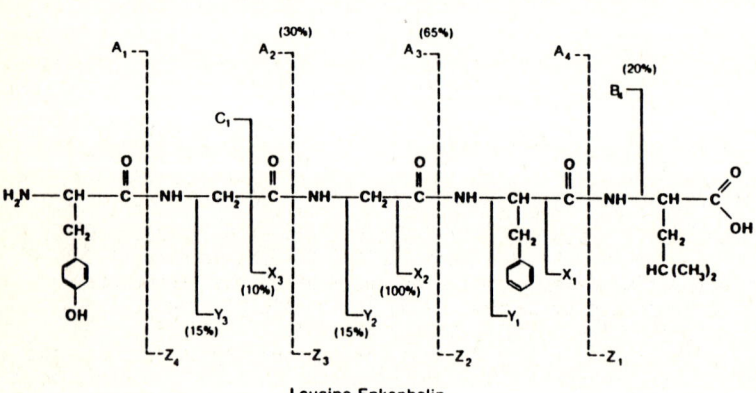

Fig. 1 The Cs^+ ion DI-FTMS spectrum of Leucine Enkephalin obtained with an incident beam energy of 5 KeV and beam current of 10×10^{-9} A cm^{-1}.

Fig. 2 Structure and proposed fragment ions of Leucine Enkephalin.

K^+, consequently $[M+Na]^+$ ions with relatively high internal energies fragment by cleavage reactions involving the organic molecule. Conversely, $[M+K]^+$ ions with relatively high internal energies dissociate to give predominantly K^+ [10].

Based on these and related studied we have demonstrated that the fragment ions observed with Cs^+ DI-FTMS can be used to obtained sequence information on small peptides. The dominant fragment ions observed in the DI-FTMS spectra of peptides (Fig. 1) correspond to cleavages C_1, A_3, and X_2 shown in Figure 2. Other fragmentation reactions which require higher energy, e.g., loss of the R substituents, are not observed because these reactions have higher energy requirements than dissociation of the $[M+Na]^+$ complex to give Na^+.

A second series of studies important to the evaluation of DI-FTMS for the analysis of large molecules are the studies on CsI [11]. In these studies we demonstrated that ions in excess of m/z 10,000 can be détected (Fig. 3). In addition, the relative abundance of the $Cs(CsI)_n^+$ cluster ions suggest that the mass discrimination associated with detection of ions by FTMS is much less than that

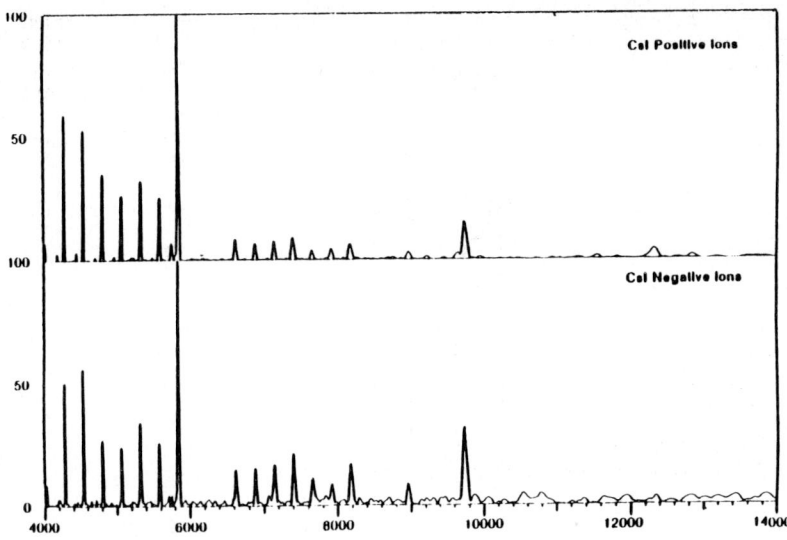

Fig. 3 Cs^+ ion DI-FTMS spectra of $Cs(CsI)_n^+$ (top) and $(CsI)_nI^-$ (bottom).

for charged particle detectors. This result is not surprising since ion detection by FTMS does not rely on the production of secondary electrons.

This study is also important because it demonstrates that the timescale (milliseconds to seconds) of the FTMS experiment does not adversely effect the data. There have been some questions raised concerning the slow decomposition reactions of large ionic species formed by desorption ionization [12], and these results suggest that the sensitivity of FTMS may be diminished by the occurrence of such slow decomposition reactions. Although the Cs^+ DI-FTMS spectra of CsI does reflect the increased timescale, note the enhancement in the "anomalous cluster ion intensities" reported by Campana [13], strong signals are observed for the larger, more stable cluster ions up to n = 62, e.g., m/z 16, 241. An additional important point arising from this study is the ability to increase the dynamic range of FTMS by the use of ion ejection methods. However, it should be noted that ion ejection techniques should be used with some caution in order to avoid collision-induced dissociation (CID) [14].

3. Conclusions

In the past several years a great deal of progress has been made in implementing high-mass FTMS. However, in the course of these studies several things have become obvious to us and other workers in this area. (1) It is possible to produce ample secondary ion yields of relatively large molecules, and detect these ions with Fourier transform methods. However, owing to the geometry and timescale of the FTMS experiment the mass spectra obtained by Cs^+ ion DI-FTMS, in particular and possibly LD-FTMS, the mass spectra obtained differ significantly from the spectrum obtained with sector, time-of-flight, and quadrupole instruments. It is clear, however, that the differences are due to the nature of the initially formed secondary ions, i.e., $[M+xNa-(x-1)H]^+$ versus $[M+H]^+$ as opposed to any inherent difference in the experiment. (2) the mass range and sensitivity of DI-FTMS is at least comparable and possibly better than that obtained with sector instrument. Also, other workers, e.g., Wilkins and Hunt, have demonstrated the high resolution capabilities of DI-FTMS. (3) Based on our work and the work of others, it is clear that the optimum experimental design for DI-FTMS is one in which the ions are generated external to the magnetic field and the ions then injected into the mass analyzer. The first approach to this was the tandem quadrupole-FTMS [7]; however, recent results by several workers demonstrate that the quadrupole is not necessary [15]. That is, a much simpler system can be designed to accomplish this process.

4. References

1. M.B. Comisarow and A.G. Marshall, Chem. Phys. Lett 25, 282 (1974)
2. C.L. Wilkins and M.L. Gross, Anal. Chem. 53, 1661A (1981)
3. R.T. McIver, Am. Lab 12, 18 (1980)
4. D.A. McGrery, E.G. Ledford and M.L. Gross, Anal. Chem. 54, 1437 (1982)
5. M.E. Castro and D.H. Russell, Anal. Chem. 56, 578 (1984)
6. C.L. Wilkins, D.A. Weil, C.L.C. Yang and C.F. Ijames, Anal. Chem. 57, 520 (1985)
7. D.F. Hunt, J. Shabanowitz, R.T. McIver Jr., R.L. Hunter and J.E.P. Syka, Anal. Chem. 57, 765-768 (1985)
8. D.F. Hunt, J. Shabanowitz, J.R. Yates III, R.T. McIver, J.E.P. Syka and J. Amy, Anal. Chem. (in press)
9. M.E. Castro, L.M. Mallis and D.H. Russell, J. Am. Chem. Soc. (in press)
10. L.M. Mallis and D.H. Russell, Anal. Chem., submitted for publication
11. M.E. Castro, D.H. Russell, S. Ghaderi, R.B. Cody, I.J. Amster and F.W. McLafferty, Anal. Chem. (in press)
12. B.T. Chait and F.H. Field, J. Am. Chem. Soc. 106, 1931 (1984)
13. T.M. Barlak, J.R. Wyatt, R.J. Colton, J.J. DeCorpo and J.E. Campana, J. Am. Chem. Soc. 104, 1212 (1982)
14. M.E. Castro and D.H. Russell, Anal. Chem. (in press)
15. P. Kofel, M. Allemann, Hp. Kellerhals and K.P. Wanczek, Int. J. Mass Spectro. and Ion Pro. 65, 97 (1985)

Application of Secondary Ion Mass Spectrometry Combined with Fourier Transform Ion Cyclotron Resonance

S. Plesko, P. Grossmann, M. Allemann, and H.P. Kellerhals

Spectrospin AG, Industriestrasse 26, CH-8117 Fällanden, Switzerland

1. Introduction

In past years, many new ionization techniques - e.g. secondary ion mass spectroscopy (SIMS), fast atom bombardment (FAB), field desorption (FD) or laser desorption (LD) - have been introduced for mass spectroscopic investigations of polar compounds, thermally labile molecules and salts.

The break-through in the popularity of FAB and liquid SIMS has been initiated by the use of a liquid matrix - usually glycerol [1]. Because the intact sample molecules can diffuse to the surface of the liquid, high primary beam intensities which are necessary for low transmission instruments can be used for long time periods without a noticeable destruction of the surface. The disadvantages of FAB (or liquid SIMS) are the occurrence of matrix effects and the high vapor pressure of glycerol. FAB also requires a vacuum system capable to handle large gas loads, making it unsuited for a direct use within an ion cyclotron resonance (ICR) instrument [2].

The use of differentially pumped external ionization sources represents one approach in an effort to use ionization techniques, which are well-established methods when a conventional mass spectrometer is applied, in the field of Fourier transform (FT) mass spectroscopy [3,4]. However, there will be some losses during the transport of ions from the source to the ICR measuring cell, and matrix effects occurring in FAB will still be present.

The use of the so-called "static SIMS" (no matrix is applied in this case) eliminates these above-mentioned problems [5]. Experiments have shown clearly that a too high intensity of the primary beam results in a very fast destruction of the sample. Obviously, the primary beam must be of such low intensity that it allows one to detect molecular or quasi-molecular ions. However, the resulting low secondary ion intensities require the use of mass spectrometers with high transmission.

The ICR mass spectrometer, capable to store nearly all ions created within the measuring cell and to analyze them simultaneously, is an ideal candidate for the use with static SIMS. Static SIMS does not deteriorate the ultra high vacuum required for FT-ICR, thus the FT-ICR spectrometers can reach the same ultra high resolution [6] and other performance characteristics in measurements on polar compounds, thermally labile molecules and salts as experienced during measurements of stable volatile compounds upon EI conditions.

2. Experimental

A standard Spectrospin CMS-47 Fourier transform mass spectrometer [7] equipped with either the standard 89mm bore or the optional 150 mm bore superconducting magnet operated at 4.7 Tesla has been used for the measurements. A special ICR measuring cell with a combined pulsed Cs^+/e^- source and corresponding power supply have been developed for static SIMS experiments. With this set-up the pulsed Cs^+ beam passes through a small hole in the front trapping plate and through the ICR cell and hits the target located in the center of the rear trapping plate. The targets are attached to the ICR cell using the solid sample inlet system.

The samples - commercially available chemicals - were used without any further purification. The usual wet preparation technique has been applied, i.e. approximately 1 ul of the sample solution (concentration about 10^{-3} mol/l to 10^{-2} mol/l) has been deposited on an etched silver surface of 7 mm dia. and dried.

The measurements have been performed using an accelerating voltage V = 2 to 4 kVolts for the primary Cs^+ ions, the peak intensity of the pulsed Cs^+ beam at the target being I = 10 nAmp, typically. The same pulse sequence as for EI measurements has been used with ionization pulse widths varying between 20 and 10 000 μsec. The pressure during the measurements reached 1 to 5 · 10^{-9} mbar.

Since secondary ion yield depends on the incident beam angle, two different geometries of the target have been used: flat targets with 0° incident beam angle and 45° targets for 45° incident beam angle, however, no systematic study has been performed.

3. Results and Discussion

The use of 45° targets improves the signal intensity roughly by a factor of 2 as compared to flat targets, however, the trapping electrostatic field in the small cell used with the 89mm bore superconducting magnet is influenced by the 45° targets and requires corrections in the plate voltages, otherwise the transient signal becomes distorted. The large ICR cell used in the 150mm bore magnet tolerates the 45° targets much better. Nevertheless, flat 0° targets have been preferred for most experiments.

A broad band static FT-SIMS spectrum of berberine chloride is shown in Fig. 1. The $(M-Cl)^+$ peak at m/z 336 is clearly the most intense signal. The presented broad band spectrum shows a full width at half-height (FWHH) resolution of m/Δm = 40 000 at m/z 336, which is limited by the acquisition time used in the present experiment. It should be kept in mind that the resolution depends on the decay time of the transient signal. Using a too short acquisition time, which is often necessary in the broad band detection mode due to the limitation in the size of computer memory, will make the instrumental resolution worse than the natural line width in the physical system. A dramatic increase of the resolution can be obtained if the narrow band detection mode is applied; e.g. in the present case the resolution improves up to m/Δm = 950 000 for the $(M-Cl)^+$ peak at m/z 336 (Fig. 2).

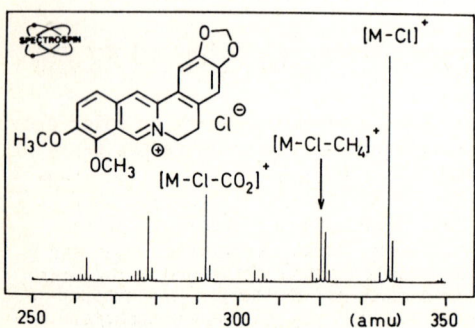

Fig. 1 FT-SIMS spectrum of Berberine chloride, positive ions

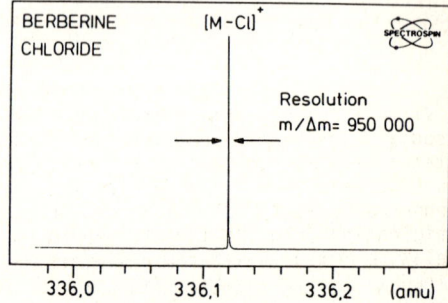

Fig. 2 High resolution FT-SIMS spectrum of Berberine chloride. FWHH resolution m/Δm = 950 000 at m/z 336

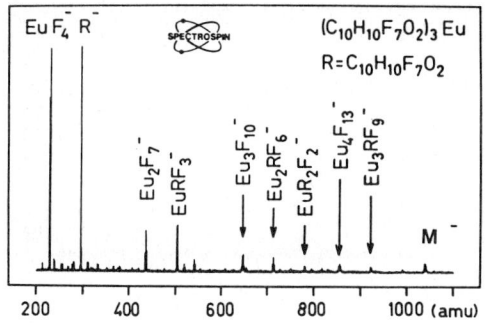

Fig. 3 FT-SIMS spectrum of Euroshift-FOD, negative ions

The utilization of the static FT-SIMS is not limited solely to salts. The broad band negative ion spectrum of the organometallic compound Euroshift - FOD $(C_{10}H_{10}F_7O_2)_3 Eu$ is presented in Fig. 3. Besides the M^- ions and the $(C_{10}H_{10}F_7O_2)^-$ peak at m/z 295, a series of cluster ions of the typ $(Eu_n R_{3n-x} F_{x+1})^-$ with $R = C_{10}H_{10}F_7O_2$ are observed. It is very likely that they are the result of ion-molecule reactions in the gas phase. The compositions of these cluster ions have been identified using the high-resolution narrow band FT mode; e.g. measurements on the molecular ions M^- (m/z 1036 and 1038) gave a resolution m/Δm = 50 000.

The cyclic peptide Gramicidin S is another test compound measured upon the above described soft ionization conditions. The quasi-molecular ions $(M+H)^+$ and $(M+Na)^+$ are observed in the broad band spectrum in Fig. 4. Using the high-resolution narrow band FT mode, a resolution of m/Δm = 19 000 has been obtained for $(M+H)^+$ peak (m/z 1142), see Fig. 5.

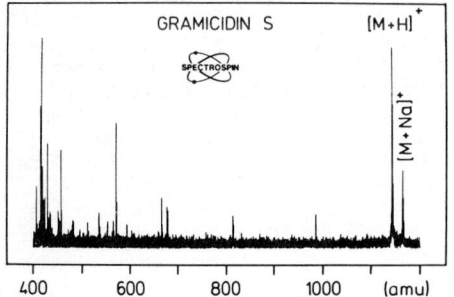

Fig. 4 FT-SIMS spectrum of Gramicidin S, positive ions

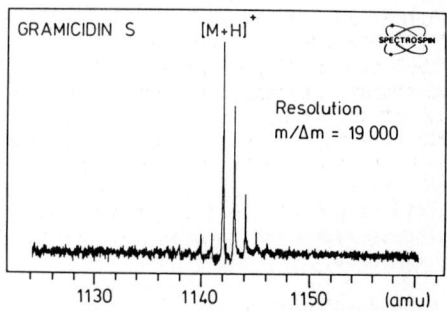

Fig. 5 High-resolution FT-SIMS spectrum of Gramicidin S, resolution m/Δm = 19 000

Vitamin B_{12} is another thermally labile non-volatile compound and frequently used as a reference compound upon soft ionization conditions. The positive ion static FT-SIMS spectrum of a sample prepared from a 10^{-2} mol/l methanol solution is displayed in Fig. 6. The low mass range (m/z < 300) contains relatively intense peaks at m/z 23 (Na^+), m/z 59 (Co^+) and silver peaks, the latter ones are originating from the target. Depending on the applied preparation procedure, the relative intensity of these signals may vary between 20% to 200% if m/z 930 is defined as the 100% peak. This FT-SIMS spectrum (Fig. 6) shows more similarities to the TOF-SIMS and ^{252}Cf plasma desorption TOF spectra [8,9] than to the first reported FT-SIMS spectrum of Vitamin B_{12} [10]. In the high mass range (m/z > 1000), the pattern of the present mass spectrum is much simpler than that of FAB mass spectrum [11]. A quite intense $(M+H)^+$ peak is present (~2%) and the intensity of

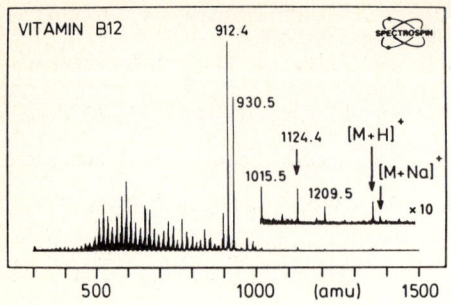

Fig. 6 FT-SIMS spectrum of Vitamin B_{12}, positive ions

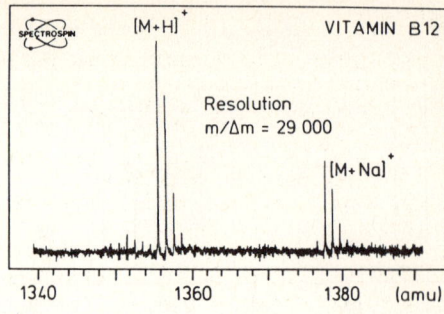

Fig. 7 High-resolution FT-SIMS spectrum of the positive quasi-molecular ions in Vitamin B_{12}, resolution $m/\Delta m = 29\,000$ at m/z 1356

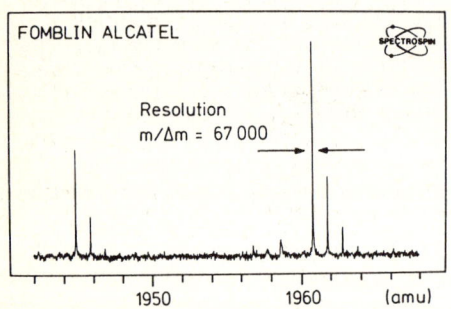

Fig. 8 High-resolution FT-SIMS spectrum of the $C_{35}F_{71}O_{12}^-$ peak (m/z 1961) of Alcatel Fomblin, $m/\Delta m = 67\,000$

the $(M+Na)^+$ varies proportional to the concentration of Na^+-ions. $(M+Cs)^+$ ions could not be observed and also the intensity of $(M-CN)^+$ peak is very weak (less than 5% of the quasi-molecular ion peak at m/z 1356). The peak at m/z 1209 can be tentatively explained by the loss of dimethylbenzimidazol, the other intense peaks find their origin likely in successive losses in the axial chain attached to the corrin ring [11]. The high resolution narrow band FT spectrum of the quasi-molecular ions is shown in Fig. 7, the resolution reaches the value $m/\Delta m = 29\,000$ at m/z 1356.

High-resolution capabilities of static FT-SIMS are further demonstrated on $C_{35}F_{71}O_{12}^-$ peak (m/z 1960.82) of Alcatel Fomblin in Fig. 8. The resolution reaches $m/\Delta m = 67\,000$.

The difference between some Fourier transform spectra and spectra taken with conventional instruments could be caused by different time scales of the experiments. In the present study, the ions are trapped for about 5 milliseconds prior to RF excitation and observation, a time which is much longer than the time scale used in TOF spectrometers and other conventional mass instruments. Chemical reactions and slow decomposition not observable in conventional spectrometry may lead to different spectra in FT mass spectroscopic studies [12].

4. Acknowledgement

The authors wish to thank Jan J. Zwinselman for his interest in this work and for many valuable discussions.

References

1. M. Barber, R.S. Bordoli, R.D. Sedgwick and A.N. Tyler: J. Chem. Soc. Chem. Commun. 1981, 325.
2. E. Onyiriuka, R.L. White, D.A. McCrery, M.L. Gross and C.L. Wilkins: Int. J. Mass Spectrom. Ion Phys. 46, 135 (1983).
3. P. Kofel, M. Allemann, Hp. Kellerhals and K.P. Wanczek: Int. J. Mass Spectrom. Ion Processes 65, 97 (1985).
4. D.F. Hunt, J. Shabanowitz, R.T. McIver Jr., R.L. Hunter and J.E.P. Syka: Anal. Chem. 57, 765 (1985).
5. A. Benninghoven: in "Ion Formation from Organic Solids", A. Benninghoven Ed.; Springer Verlag: Berlin, Heidelberg, New York, Tokyo 1983; Springer Series in Chemical Physics, Vol. 25, p. 64.
6. S. Plesko, J.J. Zwinselman, M. Allemann and Hp. Kellerhals: presented at the Meeting of the Swiss Physical Society, October -4-1984, Zürich.
7. M. Allemann, Hp. Kellerhals and K.P. Wanczek: Int. J. Mass Spectrom. Ion Phys. 46, 139 (1983).
8. P. Steffens, E. Niehuis, T. Friese and A. Benninghoven: in "Ion Formation from Organic Solids", A. Benninghoven Ed.; Springer Verlag: Berlin, Heidelberg, New York, Tokyo 1983; Springer Series in Chemical Physics, Vol. 25, p. 111.
9. W. Ens, K.G. Standing, B.T. Chait and F.H. Field: Anal. Chem. 53, 1241 (1981).
10. M.E. Castro and D.H. Russell: Anal. Chem. 56, 578 (1984).
11. M. Barber, R.S. Bordoli, R.D. Sedgwick and A.N. Tyler: Biomed. Mass Spectrom. 8, 492 (1981).
12. T. Gäumann: presented at the Meeting of the Swiss Society for Mass Spectrometry, October 1984, Rigi-Kaltbad.

Index of Contributors

Alaim, M. 142
Allemann, M. 213
Assmann, G. 79

Barofsky, D.F. 86
Barofsky, E. 86
Beavis, R. 37
Becker, O. 11
Benit, J. 164
Benninghoven, A. 56,62,67,
 74,79,96,198
Bentz, B.L. 192
le Beyec, Y. 11,42,164
Bibring, J.P. 164
Bletsos, I.V. 74
Bolbach, G. 37
Brandenberger, H. 174

Campana, J.E. 46,179
Castro, M.E. 209
Chait, B.T. 34
Chaudhary, T. 102
Chiarelli, M.P. 204
Colton, R.J. 46,51
Cooks, R.G. 28
Cotter, J.R. 11
Cotter, R.J. 142

Della-Negra, S. 11,42,164
Demirev, P. 142
Deprun, C. 164

Eicke, A. 56
Emary, W.B. 28
Ens, W. 37

Feld, H. 198
Fenselau, C. 109
Field, F.H. 102
Fife, W.K. 28
Freas, R.B. 179

Greifendorf, D. 67
Gross, M.L. 134,204

Grossmann, P. 213
Grotjahn, L. 118
Guthier, W. 11,17

Hayashi, A. 113
Hedin, A. 6
Heller, T. 198
Hercules, D.M. 74,147
Hillenkamp, F. 153
Holtkamp, D. 62,153
Honig, R.E. 192
Honovich, J. 142
Hsu, B.-H. 28
Hyver, K. 109
Håkansson, P. 6

Jabs, H.-U. 79
Jonsson, G. 6
Junack, M. 96
Jungclas, H. 22

Karas, M. 153
Katakuse, I. 113
Katz, R.N. 102
Kellerhals, H.P. 213
Kempken, M. 62
Kidwell, D.A. 46,51
Kissel, J. 169
Klüsener, P. 62,153
Krueger, F.R. 169

Lange, W. 67
van Leyen, D. 67,74
Lindner, B. 158

Macfarlane, R.D. 2
Main, D. 37
Marien, J. 103
Matsuda, H. 113
Matsuo, T. 113
Mayer, F.J. 169
McCrery, D.A. 204
McNeal, C.J. 2

Natalis, P. 103
Niehuis, E. 67,74,198

Olthoff, J. 142

de Pauw, E. 103
Pelzer, G. 103
Phelps, R.G. 2
Plesko, S. 213

Röllgen, F.W. 91
Renner, D. 126
Rocard, F. 164
Roepstorff, P. 6
Ross, M.M. 46,51
Russell, D.H. 209

Säve, G. 6
Sakurai, T. 113
Salehpour, M. 6
Schmidt, L. 22
Schueler, B. 37
Seydel, U. 158
Sichtermann, W. 96
Spiteller, G. 126
Standing, K.G. 37
Sundqvist, B. 6

Tomer, K.B. 134

Wada, Y. 113
Walter, M. 79
West, F.B.Ch. 174
Westmore, J.B. 37
Widdiyasekera, S. 6
Wien, K. 11
Wirth, K.P. 91
Wollnik, H. 184
Wong, S.S. 91

Springer Series in Surface Sciences

Editors: G. Ertl, R. Gomer

Volume 1
H. J. Kreuzer, Z. W. Gortel
Physisorption Kinetics
1986. 133 figures. XIII, 325 pages
ISBN 3-540-16176-7

Contents: Introduction. – Gas-Solid Interaction. – The Master Equation. – Transition Probabilities in the Master Equation. – Desorption Times. – Time of Flight Spectra. – Sticking and Accomodation. – Kramers Equation. – Summary and Outlook. – References. – Subject Index.

Volume 2
The Structure of Surfaces
Editors: M. A. van Hove, S. Y. Tong
1985. 223 figures. XII, 435 pages
ISBN 3-540-15410-8

Contents: Introduction. – Theory of Surface Structure: General Discussion. Specific Applications. – New Surface Structure Techniques: Techniques Based on Electrons. – Developments in Existing Techniques: LEED and Electron Propagation. Diffuse LEED, NEXAFS/XANES and SEXAFS. High-Resolution Electron Energy Loss Spectroscopy (HREELS). Atom and Ion Scattering. Photoemission. Neutron Scattering. – Clean and Adsorbate-Covered Metals: Clean Metal Surfaces. Atomic Adsorption on Metal Surfaces. Molecular Adsorption on Metal Surfaces. – Semiconductors: Elemental Semiconductors. Compound Semiconductors. Adsorbate-Covered Semiconductors. – Defects and Phase Transitions: Theoretical Aspects. Experimental Studies. – Index of Contributors. – Subject Index.

Volume 3
Dynamical Phenomena at Surfaces, Interfaces and Superlattices
Proceedings of an International Summer School at the Ettore Majorana Centre, Erice, Italy, July 1–13, 1984
Editors: F. Nizzoli, K.-H. Rieder, R. F. Willis
1985. 194 figures. XIII, 329 pages
ISBN 3-540-15505-8

Volume 4
Desorption Induced by Electronic Transitions DIET II
Proceedings of the Second International Workshop, Schloß Elmau, Bavaria, October 15–17, 1984
Editors: W. Brenig, D. Menzel
1985. 164 figures. IX, 291 pages
ISBN 3-540-15593-7

Thin-Film and Depth-Profile Analysis
Editor: H. Oechsner
With contributions by H. W. Etzkorn, W. O. Hofer, S. Hofmann, J. Kempf, J. Kirschner, U. Littmark, H. J. Mathieu, H. Oechsner, J. M. Sanz, H. Wagner, H. W. Werner
1984. 99 figures. XI, 205 pages. (Topics in Current Physics, Volume 37). ISBN 3-540-13320-8

Contents: Introduction. – The Application of Beam and Diffraction Techniques to Thin-Film and Surface Micro Analysis. – Depth-Profile and Interface Analysis of Thin Films by AES and XPS. – Secondary Neutral Mass Spectrometry (SNMS) and Its Application to Depth-Profile and Interface Analysis. – In-Situ Laser Measurements of Sputter Rates During SIMS/AES In-Depth Profiling. – Physical Limitations to Sputter Profiling at Interfaces – Model Experiments with Ge/Si Using KARMA. – Depth Resolution and Quantitative Evaluation of AES Sputtering Profiles. – The Theory of Recoil Mixing in Solids. – Additional References with Titles. – Subject Index.

Ion Formation from Organic Solids
Proceedings of the Second International Conference, Münster, Federal Republic of Germany, September 7–9, 1982
Editor: A. Benninghoven
1983. 170 figures. IX, 269 pages. (Springer Series in Chemical Physics, Volume 25). ISBN 3-540-12244-3

The proceedings deals with the ion formation involatile, thermally labile organic compounds. Fundamental aspects of the ion formation process as well as present applications and future trends in the analytical application are treated. In addition, instrumental as well as technological developments as time of flight instruments or liquid matrices, are described. This book will provide the reader with an authoritative guide for analytical work in the life sciences that will retain its validity for many years to come.

Springer-Verlag Berlin Heidelberg New York Tokyo

Springer Series in Chemical Physics

Editors: V.I. Goldanskii, R. Gomer, F.P. Schäfer, J.P. Toennies

A Selection:

Volume 9
Secondary Ion Mass Spectrometry SIMS-II

Proceedings of the Second International Conference on Secondary Ion Mass Spectrometry (SIMS II)
Stanford University, Stanford, California, USA, August 27–31, 1979
Editors: A. Benninghoven, C. A. Evans, Jr., R. A. Powell, R. Shimizu, H. A. Storms
1979. 234 figures, 21 tables. XIII, 298 pages
ISBN 3-540-09843-7

Contents: Fundamentals. – Quantitation. – Semiconductors. – Static SIMS. – Metallurgy. – Instrumentation. – Geology. – Panel Discussion. – Biology. – Combined Techniques. – Postdeadline Papers.

Volume 19
Secondary Ion Mass Spectrometry SIMS III

Proceedings of the Third International Conference, Technical University, Budapest, Hungary, August 30–September 5, 1981
Editors: A. Benninghoven, J. Giber, J. László, M. Riedel, H. W. Werner
1982. 289 figures. XI, 444 pages
ISBN 3-540-11372-X

Contents: Instrumentation. – Fundamentals I. Ion Formation. – Fundamentals II. Depth Profiling. – Quantification. – Application I. Depth Profiling. – Application II. Surface Studies, Ion Microscopy. – Index of Contributors.

Volume 36
Secondary Ion Mass Spectrometry SIMS IV

Proceedings of the Fourth International Conference, Osaka, Japan, November 13–19, 1983
Editors: A. Benninghoven, J. Okano, R. Shimizu, H. W. Werner
1984. 415 figures. XV, 503 pages
ISBN 3-540-13316-X

Contents: Fundamentals. – Quantification. – Instrumentation. – Combined and Stataic SIMS. – Application to Semiconductor and Depth Profiling. – Organic SIMS. – Application: Metallic and Inorganic Materials. Geology. Biology. – Index of Contributors.

Volume 44
Secondary Ion Mass Spectrometry SIMS V

Proceedings of the Fifth International Conference, Washington, DC, September 30–October 4, 1985
Editors: A. Benninghoven, R. J. Colton, D. S. Simons, H. W. Werner
1986. 388 figures. XXI, 561 pages
ISBN 3-540-16263-1

Contents: Retrospective. – Fundamentals. – Symposium: Detection of Sputtered Neutrals. – Detection Limits and Quantification. – Instrumentation. – Techniques Closely Related to SIMS. – Combined Techniques and Surface Studies. – Ion Microscopy and Image Analysis. – Depth Profiling and Semiconductor Applications. – Metallurgical Applications. – Biological Applications. – Geological Applications. – Symposium: Particle-Induced Emission from Organics. – Organic Applications and Fast Atom Bombardment Mass Spectrometry. – Index of Contributors.

Springer-Verlag
Berlin Heidelberg New York Tokyo

RAYMOND H. FOGLER LIBRARY
DATE DUE

BOOKS ARE SUBJECT TO
RECALL AFTER TWO WEEKS

APR 15 1987